Visualised Systems Engineering on Railway Projects

by Jong-Pil Nam

AuthorHouse™ UK
1663 Liberty Drive
Bloomington, IN 47403 USA
www.authorhouse.co.uk
UK TFN: 0800 0148641 (Toll Free inside the UK)
UK Local: 02036 956322 (+44 20 3695 6322 from outside the UK)

This book is printed on acid-free paper.

ISBN: 979-8-8230-9117-6 (sc)
ISBN: 979-8-8230-9118-3 (e)

Library of Congress Control Number: 2024926126

Print information available on the last page.

Published by AuthorHouse 12/19/2024

authorHOUSE®

Preface

Systems Engineering exists to integrate engineering objectives, processes and activities in various industries especially in railway projects since railway infrastructure consists of multiple disciplines. Thus, lots of engineers who are involved in systems engineering are looking for applicable standards and regulations. For subcontractors or suppliers providing small equipment, it is relatively easy to get applicable guidelines such as ISO standards or EU standards. However, it is sometimes difficult for the main contractor at the top level in an organisational structure in railway projects to apply Systems Engineering methodologies as a system integrator. Of course, there are many textbooks and guidelines like INCOSE or SEBok that introduce the systems engineering process, however, it is not easy for systems integrators to choose any specific process to apply because those guidelines have a lot of contents to cover most industries. And furthermore, it is necessary modify and apply such methodologies to suit the characteristics of the railway projects.

My colleagues also have faced the same problems, and I sometimes felt the need to help them. So, I have developed a lot of processes that explain how to conduct systems engineering tasks. Using those graphical materials, I held several training sessions. While performing the sessions, I have prepared more slides, and these materials were very helpful to anyone who is struggling with certain issues in their systems engineering work. This is the reason why I have prepared this book.

This book focuses on the fundamental and practical applications of systems engineering to railway projects, and it covers most areas of System Engineering that systems engineers should know such as railway operation & performance, requirements, V&V, RAMS, configuration, cyber security, human factors, etc. This book introduces Systems Engineering considering the relationships between clients, contractors, and sub-contractors because the relationships between them affect the activities in the process of systems engineering.

The illustrations and graphical materials in this book will help readers easily understand the contents of this book. I'm sure you will gain an insight into systems engineering if you read this book.

I wish you success with your railway projects.

Jong-Pil Nam
M.Sc. of Railway Systems Engineering, PMP (Project Management Professional)
kity0607@gmail.com

About the Author

Jong-Pil Nam

He began his career at Korea Rail Network Authority in South Korea in 1995. and between 2011 and 2012, he studied for a master's degree in Railway Systems Engineering and Integration at the University of Birmingham, UK. He won the Best Technical Dissertation from the University of Birmingham for his dissertation.

As a system engineer or a system integrator, he has participated in lots of railway projects – Western Sydney Airport (Australia), MRT-7 (Philippines), LRT1 (Indonesia), GTX-A (South Korea), National Rail Network (Saudi Arabia), Korea high-speed rail (South Korea), Bundang Metro (South Korea), etc.

With his academic careers and vast experience of systems engineering in railway projects, he sometimes gives lectures on Systems Engineering while participating in several Systems Engineering projects. He prepared this book based on his railway experiences and academic careers to help railway Systems Engineering beginners.

Table of contents

OVERVIEW OF SYSTEMS ENGINEERING

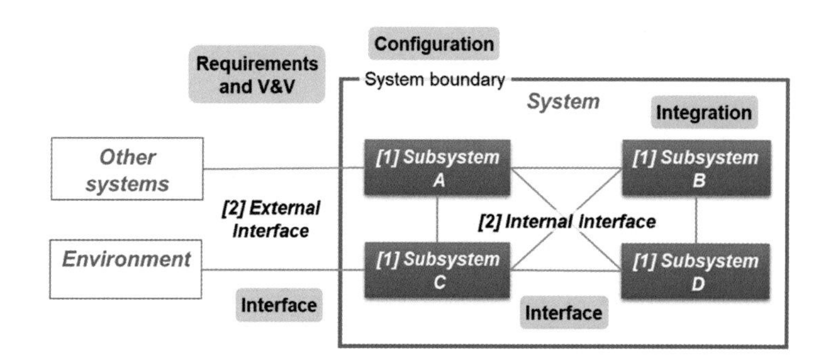

What is Systems Engineering?

Railway infrastructure is largely composed of structural elements and E&M (Electrical and Mechanical) systems. A structural element refers to fixed elements such as building pillars and walls that provide foundational support and stability. Conversely, E&M systems refer to devices or equipment that ensure the effective functioning of the railway. These systems are crucial for the operational aspects of the railway. Since structural components are fixed facilities, once installed, they do not require ongoing consideration of operational issues. In contrast, E&M systems must operate correctly, effectively, and safely. Thus, systems engineering focuses on E&M systems related to these operational aspects.

Properties of a system

To gain a profound insight into Systems Engineering, it is important first to comprehend the properties of a system. From an industrial perspective, a system can be defined as [1] an integrated set of components [2] that interact with each other and the environment [3] to perform the required functions and achieve the desired performance [4] while minimising failures or risks. The properties of a system can be described as follows:

[1] A system is an integrated set of components.
[2] Components of the system interact with each other and the environment.
[3] The system should perform the required functions and achieve the desired performance.
[4] The system should operate with minimised failures and risks.

When it comes to [1], a system consists of multiple subsystems or components classified as actuators, controllers, and monitors. System development involves numerous suppliers (or subcontractors) that contribute to the development of different components, which must be properly integrated. The integration process requires meticulous planning and coordination to ensure that all parts work together harmoniously. Regarding [2], interface issues frequently arise both between subsystems and between the system and its environment, which requires careful coordination and integration to ensure seamless operation. Moreover, system developers must consider the environmental impact of the system they are creating.
[3] represents the ultimate purpose of a system. Throughout the system development lifecycle, it is crucial to meticulously design and produce each component to meet the required functions and performance.

Additionally, the integrated system, formed by combining these components, must also meet the required functions and performance.
When it comes to [4], risk management is a critical aspect to consider when designing a system. Figure 1 shows the types of risks and failure.

Figure 1 - Risk types and causual factors

Risk type	Category of source	Causal factor	Example
Systematic failures	System itself	Poorly designed system without function to prevent hazards or having malfunction to cause hazards	Train doors with hard and sharp edges
Random failures	System itself	Inherences failure rate of equipment	Irregular failure of a device
Human errors	Operators, maintainers, and passengers	Incorrect use by operators, maintainers, and passengers	A driver in a train cab presses a wrong button by mistake
Cyber security failures (IT security)	Attackers outside the system	An attack via IT network	Railway signalling disturbance due to hacking signalling systems
Physical security failures	Attackers outside the system	An attack on railway infrastructure and systems	Access of unauthorised person to OCC (Operating Control Centre)
Intended failures (SOTIF-related)	Surrounding environment	Technological limitation for unforeseen situations and scenarios	A driving accident involving a self-driving car unable to deal with unforeseen situations

< Product Breakdown Structure (PBS) >
In this book, the system hierarchy is defined as follows: the top level of the system hierarchy is referred to as the 'System level,' which consists of subsystems. Below this, the 'Subsystem level' consists of components. In the context of railway systems, the levels of composition are as follows:

- System level: a railway line.
- Subsystem level: track, signalling, power supply, communication, rolling stock, etc.
- Component level: the individual elements of subsystems.

The term 'equipment' is also used in this book. When referring to human-machine equipment, it is assumed that the equipment consists of actuator, controller, and monitor.

Systems Engineering

Systems Engineering is a methodical approach that guides engineers in designing and developing products with optimised cost and performance. It focuses on the entire lifecycle of a complex system,

ensuring that it is designed, produced, and managed effectively to meet all system requirements. By offering a structured framework, Systems Engineering enables engineers to enhance their products and achieve the desired functions and performance.
At the subsystem level, subsystem engineers participate in the Systems Engineering process, resolving interface and interference issues with other subsystems. They also need to reflect the allocated functions and performance in their design. Failure to address the issues can lead to a lack of consistency, non-compliance, and other problems.

Since all systems engineering (SE) activities aim to meet the project requirements, the level of SE is dependent on the complexity of the requirements. In this book, the author defines high-level requirements as "difficult and complex requirements to fulfil." Increasing the level of project requirements as much as possible will significantly enhance system quality, but that also leads to increased project costs. Conversely, the higher the level of requirements, the lower the failure cost of the system (or product). Therefore, it is important to find an optimal level of requirements, as shown in Figure 2.

Figure 2 - Optimal requirements level

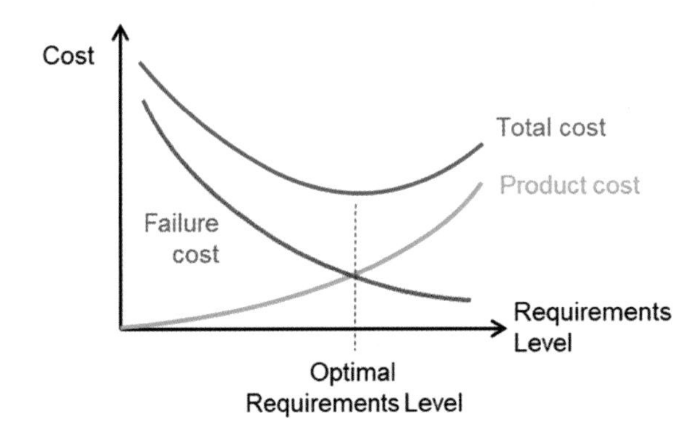

System Assurance

System assurance is frequently addressed in systems engineering activities. What does 'assured system' mean? An assured system provides a high degree of confidence that the system is expected to perform its intended function under stated conditions.
System Assurance refers to a justified level of confidence that an integrated system, comprising a collection of subsystems, will meet the requirements in an integrated manner. Thus, the activities of

System Assurance are the processes to meet the system requirements. For successful systems engineering work, the System Assurance process is required. Similarly, to achieve System Assurance, systems engineering methodologies should be implemented.
An assured system must be processed at least as follows:

- For equipment, all materials used in equipment development must be certified or must pass relevant tests.
- Each component must be verified and validated through all phases, from planning phase to T&C (Testing and Commissioning) phase.
- In the final phase of development, the final integrated system must be tested for client acceptance.
- When applicable, the quality of the final deliverables must be demonstrated during the demonstration period while operating.

To assure a system, the following procedures are required:

- Requirements must be complete and clearly defined.
- Design and development must be carried out correctly.
- Tests must be appropriately planned and rigorously conducted.
- Tests must be conducted by qualified people who have no conflict of interest.
- The quantity of supporting evidence must be sufficient.

The framework of system assurance should be applied to systems engineering activities to successfully develop a system, and these activities should be planned in a Systems Engineering Management Plan (SEMP). A Systems Engineering Management Plan provides a structured approach to managing and executing the systems engineering process throughout a project lifecycle. It defines roles, responsibilities, processes, tools, and timelines, ensuring that all stakeholders are aligned and that system requirements are met efficiently. Additionally, it helps minimise risks and ensures quality control throughout the project lifecycle.
A Systems Engineering Management Plan should include the following contents:

- Definitions and references: definitions of terms, abbreviations, and relevant references.
- System overview.
- Systems engineering overview: roles of systems engineering, definition of lifecycle, and lifecycle methodology.
- Systems engineering management organisation: scope and responsibilities of each party.

- Components of systems engineering activities: requirements management, interface management, configuration management, RAM management, safety management, human factors integration, verification & validation, integration management, and more.
- Information management: communication processes, record-keeping, and approval procedures.

Areas of Systems Engineering in railway projects

Depending on the type of project or system, Systems Engineering tasks in railway projects may consist of the following components:

- SE Management – Manage the systems engineering processes, internal schedule, and resources.
- Preliminary O&M (Operation and Maintenance) Planning – Establish the preliminary O&M plan.
- Operation Performance – Control the design of infrastructure and systems to assist engineers in each discipline in achieving operational targets, such as RTT (Round Trip Time).
- Requirements – Manage system requirements.
- V&V (Verification and Validation) – Verify and validate deliverables and activities of system suppliers to ensure they meet the system requirements.
- Configuration – Control the project or systems configurations.
- Interface – Manage interface items and issues between subsystems, equipment, products, and devices.
- Software – Manage the software development process to ensure compliance with software requirements.
- RAM – Manage the reliability and maintainability of each subsystem, equipment, product, and device to achieve target availability.
- Safety – Manage system safety using safety methodologies.
- Human Factors – Manage system design to ensure the man-machine interface is fit for use.
- EMC (Electro-Magnetic Compatibility) – Manage the design of subsystems, equipment, products, and devices to eliminate or minimise EMC-related issues
- Noise & Vibration – Eliminate or minimise noise and vibration caused by train operations.
- Cyber security – Manage the design of networks and software to ability to defend against cyber-attacks.
- T&C (Testing and Commissioning) – Manage T&C processes and activities.
- Integration – Manage requirements allocation and interface among subsystems, equipment, products, and devices to ensure smooth aggregation.
- RAM demonstration (after opening) – Detect, record, and analyse failures to manage failures occurring during the operation phase.

Of course, the disciplines of systems engineering vary greatly depending on project characteristics and size. They function like modules, so the scope of SE can selectively include various components, such as EMC, cybersecurity, and others.

Based on the author's experiences, the framework of Systems Engineering activities can be outlined as shown in Figure 3.

Figure 3 – Framework of Systems Engineering

Engineering Management in Development Phase with ISO 9001 and ISO 55001

Systems Engineering Framework

Control proces

Architecture

Integration

Interface

Configuration

Change Control Board

Requirements management (define, decompose, and allocate)

Rail domain requirements (function & performance)

Signalling

Power supply

Communication

Rolling stock

Depot

Track

Station

Trackbed (civil)

Rail Systems Engineering requirements

Oper. Performance

RAM

Safety

Human Factors

EMC

Software

Cybersecurity

Noise & Vibration

Deliverables management (actualise, test, and integrate)

Designing

Manufacturing

Installing

Testing & Commissioning

Verification & Validation

Input

Verification & Validation

Input

Traceability Matrix (Requirements Database)

In a project, there will be four groups for systems engineering activities as follows:

- [1] Control process: this process intervenes in and coordinates with all engineering and systems engineering activities.

- [2] Requirements management: all requirements of rail domains and systems engineering will be defined, decomposed, allocated to each team, and managed through the requirements management process by using a requirements database.
- [3] Deliverables management: all requirements are actualised (designed, manufactured, installed, tested, and integrated) according to requirements specifications (decomposed and allocated requirements).
- [4] All engineering results and deliverables from [2] and [3] must be verified and validated through the V&V process to ensure that engineering results and deliverables meet the requirements specifications. And then the results of the V&V will be stored in the traceability matrix (or requirements database).

Near the end of the project, a compliance matrix will be prepared by capturing the original requirement descriptions from the contractual documents and referring to the evidence in the traceability matrix (or requirements database). Based on the compliance matrix, the deliverables are checked and accepted by the client. ISO 9001 and ISO 55001 provide guidance for establishing the framework, and all systems engineering activities are governed by the standards.

System Integration and architecture

Systems integration is the process of combining different subsystems or components into a single, unified system. It involves ensuring that these components work together seamlessly to achieve the overall functionality and performance.

The activities of systems integration can be summarised as below:

- Completely assemble the implemented components to ensure compatibility.
- Demonstrate that the integrated system performs the expected functions and meets the requirements relating to performance and effectiveness.
- Detect defects and faults in design and assembly activities during the V&V (Verification and Validation) process

This process primarily includes system architecting, requirements management, configuration management, interface management and V&V (Verification and Validation) activities. The goal of system integration is to create a cohesive system that operates efficiently and effectively as a whole. When it comes to actual Systems Engineering (SE) services, the terms 'project manager' of the SE service, 'system integrator,' and 'system engineer' can be used interchangeably.

As a system is described in the previous chapter, as follows:

[1] A system is an integrated set of components.
[2] Components of a system interact with each other and the environment.
[3] The system should perform the required functions and achieve the desired performance.
[4] The system should operate with minimised failures and risks.

Each part of Systems Engineering and Integration will play the following roles:

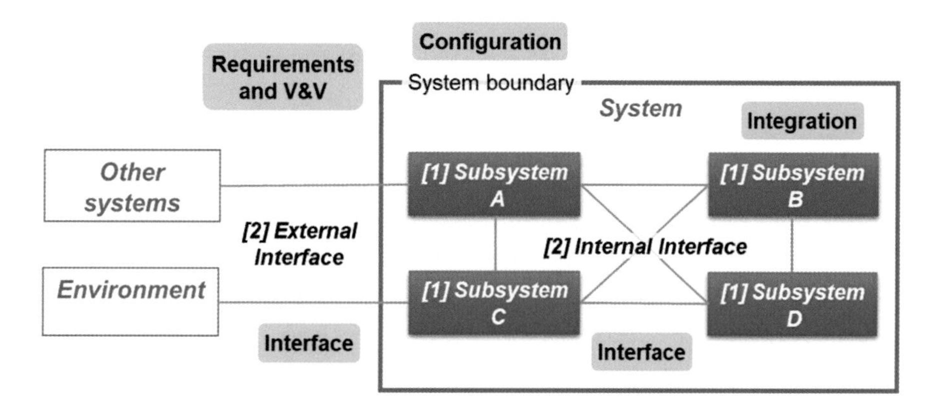

- Integration management, requirements management and configuration management help define the system boundary and system components.
- The interface management process aims to eliminate interface issues between system components and issues in the interface with other systems and the environment.
- Generally, project requirements include interface items.

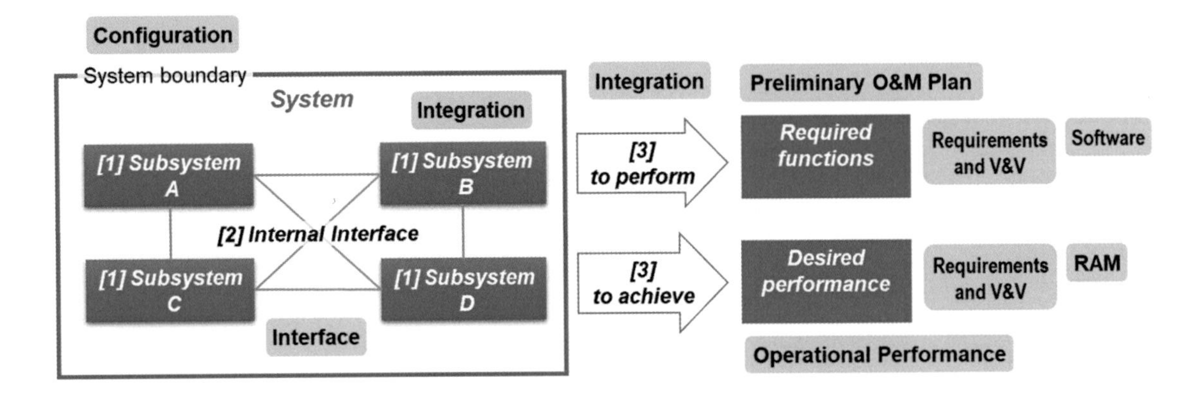

- The Preliminary O&M Plan, requirements management activities, including V&V activities, and software assurance process contribute to defining the required functions and ensuring the system performs them.
- Operational performance, requirements management process, and RAM management activities help define the target performance and ensure that the system achieves it.
- Integration management process contributes to the allocation of sub-function and sub-performance requirements to each system component.
- Management of safety, interface, software, EMC, and noise & vibration helps to minimise systematic failures in a system.
- RAMS management contributes to minimise random failures in a system.

[4] With minimised failures and risks **Requirements and V&V**

Risk type	Category of source	Causal factor	
Systematic failures	System itself	Poorly designed system without function to prevent hazards or having malfunction to cause hazards	**Safety, interface, software, EMC, N&V**
Random failures	System itself	Inherences failure rate of equipment	**RAM, safety**
Human errors	Operators, maintainers, and passengers	Incorrect use by operators, maintainers, and passengers	**Human Factors**
Cyber security failures (IT security)	Attackers outside the system	An attack via IT network	**Cybersecurity**
Physical security failures	Attackers outside the system	An attack on railway infrastructure and systems	
Intended failures (SOTIF-related)	Surrounding environment	Technological limitation for unforeseen situations and scenarios	

- The human factors integration process can minimise human errors among operators, maintainers, and passengers when operating, maintaining, and using a system, respectively.
- Cybersecurity management reduces opportunities for cyber-attacks on the system via the IT network.

System architecture refers to the design and organisation of a complex system. It defines how various components operate and interact with each other. Here is a simplified breakdown:

- Components: the individual parts of the system, such as hardware, software, and data.
- Interactions: how components interact and function together.
- Layers: different levels of the system, such as application, middleware, and hardware.
- Configuration: the setup and organisation of components within the system.
- Design principles: guidelines for designing the system, including reusability, scalability, flexibility, and reliability.
- Documentation: diagrams and descriptions explaining the system's structure and functionality.

Relationship between SE and certificates

Some people may be confused about the relationships between Systems Engineering and certificates (ISA, ASBO, NOBO, and DEBO). Each certificate simply means:

- ISA (Independent Safety Assessment): Certificate ensuring the safety of products and systems.
- ASBO (Assessment Body): Certificate granted based on international standards and regulations such as ISO standards.
- NOBO (Notified Body): EU certificate based on EU standards and regulations, such as EU standards.
- DEBO (Designated Body): Domestic certificate based on domestic standards and regulations.

To issue a certificate, an assessment team must verify and validate a product or system by reviewing, assessing, and analysing. The results of Systems Engineering activities or deliverables processed by Systems Engineering framework, will be assessed by the assessment team.

Application of Systems Engineering on railway projects

Typical system development activities

Engineers need to understand typical system development activities to ensure the successful integration of all system components, manage technical risks, and make informed decisions throughout the project lifecycle. This knowledge helps coordinate efficiently with other teams, ensure that system requirements are met, and ensure alignment across design, development, testing, and implementation.

Typical system development activities can include:

- Concept Development: Developing multiple alternative concepts to meet the system requirements, and selection of the most promising concept.
- Requirements Analysis: The process of defining and documenting the system requirements based on customer needs, stakeholder expectations, and operational requirements.
- Design: Transforming the selected concept into a detailed system and component design of the system and its components.
- Configuration Management: The process of controlling and tracking the changes to the system and its components over its life cycle.
- Actualisation: Manufacturing and installing the system based on the approved design.
- Verification and Validation (V&V): The process of evaluating the system and its components to ensure that they meet the specified requirements.
- Integration and Test: The process of bringing together the system components, verifying that they work together as intended, and validating that the system meets the specified requirements.
- Operations and Maintenance: The processes and activities required to operate, maintain, and support the system over its life cycle.
- Risk Management: The process of identifying, assessing, and mitigating risks associated with the system and its components.
- Decision Making: Making informed decisions based on systems engineering results and the analysis of trade-offs between cost, schedule, and performance.
- Demonstration: Proving system quality or availability.

V-cycle process

V-cycle process is a process that represents the Systems Engineering (SE) process. EN 50126, a standard used world-widely for railway SE, suggests V-cycle process as shown in Figure 4.

Figure 4 – V-Cycle process

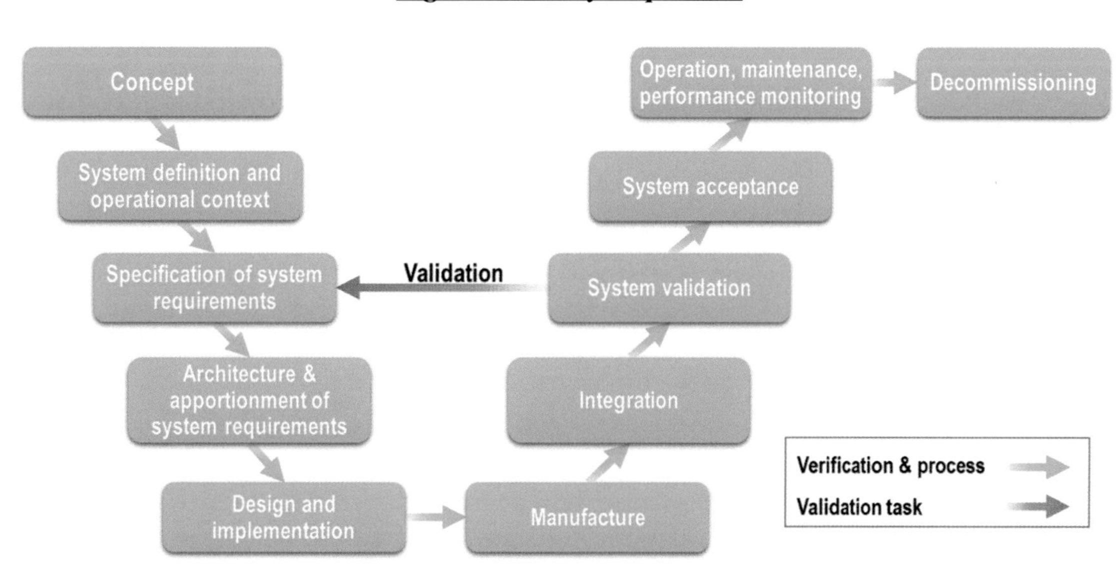

- Concept – this phase helps to understand the system to be developed.
- System definition and operational context – the objective of this phase is to define operational requirements, the system, and the system boundary.
- Risk analysis and evaluation – hazards are defined, and risks are assessed in this phase.
- Specification of system requirements – this phase specifies overall system requirements.
- Architecture and apportionment of system requirements – the objective of this phase is to apportion system requirements to the subsystems.
- Design and implementation – in this phase, subsystems are designed.
- Manufacture – subsystems and components are manufactured in this phase.
- Integration – in this phase, subsystems are assembled and installed, and the integrated system is demonstrated to works as defined requirements.

- System validation – the objective of this phase is to confirm, through examination and provision of evidence, that the system complies with the requirements.
- System acceptance – the objective of this phase is to assess the compliance of the system with the overall requirements.
- Operation, maintenance, and performance monitoring – in this phase, the system is operated, maintained, and monitored.
- Decommissioning – the objectives of this phase are to decommission the system.

However, the above process has been established from the RAMS perspective. Other referable standards or handbook, such as the INCOSE Systems Engineering Handbook, SEBOK (Systems Engineering Body of Knowledge), suggest slightly different processes and steps.
In some non-European countries, the system development process is regulated by domestic laws. Thus, engineers in those countries may find it difficult to faithfully follow the V-process.

Systems engineering processes can be simply categorised into three phases: decomposition, actualisation, and integration phase, as shown in Figure 5.

Figure 5 - V-Cycle

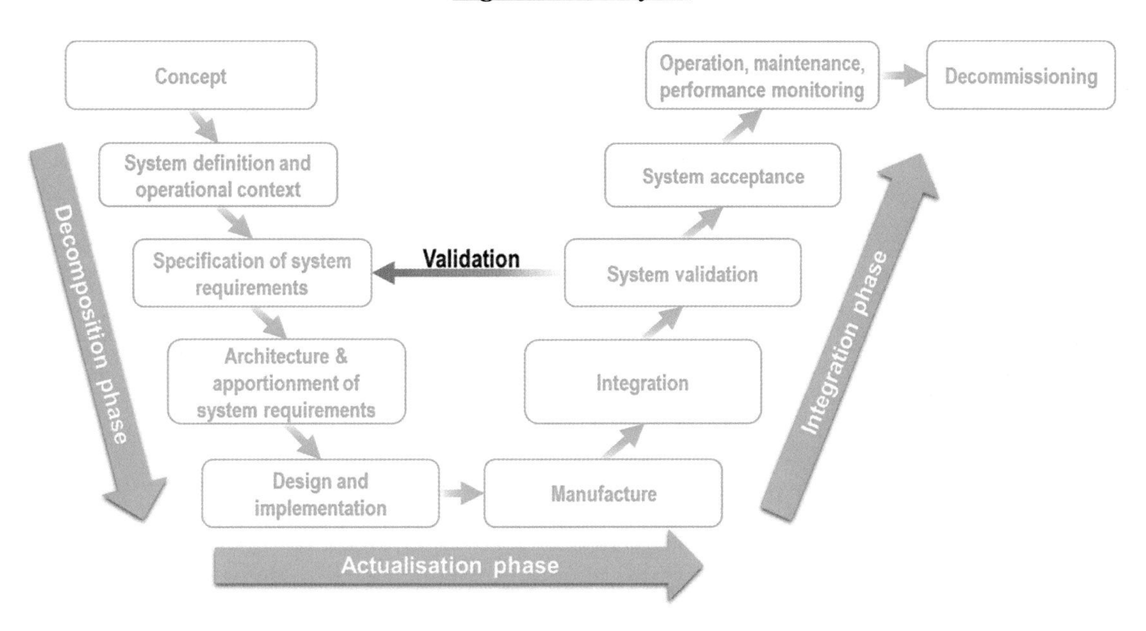

Systems Engineering in railway projects

Even though EN 50126 and other SE-related standards suggest techniques and methodologies for Systems Engineering, it is not easy for the original contractor at the top level of the organisational structure in railway projects to apply these standards. The project owner, permitted by the government for the railway project, must comply with all railway-related laws and regulations. Therefore, anyone who wants to apply the Systems Engineering process to their project must first study the applicability of the V process of Systems Engineering to their projects, as the methodologies of SE introduced in the standards may conflict with the railway-related legal requirements. At the lower level of systems, where there are no legal requirements but technical requirements, engineers, such as signalling or power supply engineer, can design based on EN 50126.

Figure 6 – Level of organisation

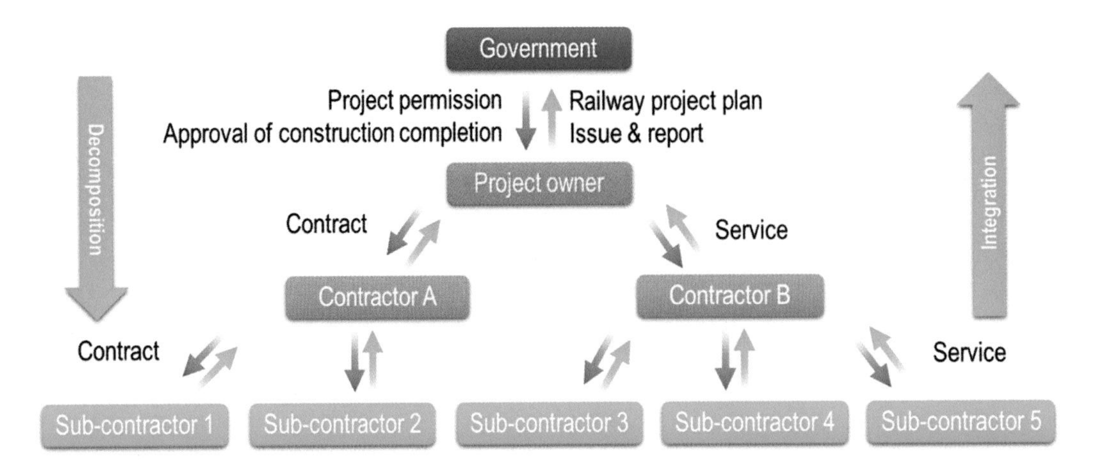

Figure 6 outlines typical structure of organisations for railway projects. The project owner, who is permitted by the government to execute railway projects, must comply with all railway-related laws and regulations. The government reviews and approves the plan if it complies with applicable laws, regulations, and governmental master plan for railways. Organisations at the higher levels are subject to laws and regulations, while those at the lower levels are governed by technical standards.

The government that permits railway projects can be either a central government or a local authority. The project owner executes the plan by contracting with a design company and a construction company, who then award contracts to subcontractors. In this book, the top level in the organisational structure refers to the original (or principal) contractor.

Figure 7 - V-Cycle process of national railway projects

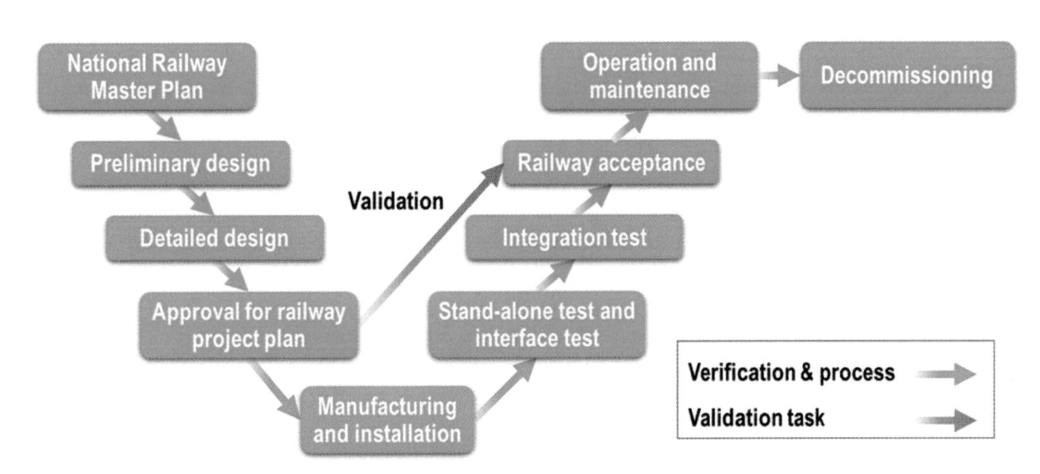

Figure 7 shows the V-cycle process for national railway construction projects in South Korea, based on railway-related laws and regulations.

- National Railway Master Plan – Based on regional population and transportation demand analysis, the government establishes the National Railway Master Plan for building national railways. Local governments follow a similar process for constructing metro lines.
- Preliminary design – The preliminary design is conducted to define the scope of the railway project and estimate the project budget.
- Detailed design – Based on the preliminary design and budget estimation, the detailed design is developed.
- Approval for railway project plan – Based on detailed design and budget estimation, the government approves the railway project plan.
- Manufacturing and installation – After government approval, manufacturing and installation are implemented.
- Stand-alone test and interface test – The stand-alone test is performed by each contractor on their own subsystems and the interface test is performed to ensure that the different subsystems can communicate and work with each other effectively.
- Integration test – The goal of integration testing is to ensure that the different components of the system work together as intended and that the system meets the specified requirements.

- Railway acceptance – After all railway components are verified and validated through stand-alone test, interface test, and integration test, government accepts the railway.
- Operation and maintenance – Korail, a Korean national rail operator, takes over to operate and to maintain the railway.
- Decommissioning – It is the process of retiring or shutting down a system or component at the end of its useful lifecycle.

Organisational structure and PBS

Size of a railway project is generally huge and there are many types of railway project implementation in terms of contract. In each type of project implementation, the project owner may have the following types of contractors:

- Turn-key (TK) contract – the TK contractor develops the project concept and performance requirements during the bidding phase and implements design and construction.
- Design and build (DB) contract – the client develops the project concept and performance requirements by hiring a design company, and the DB contractor implements detailed design and construction.
- Separated order – the client develops the project concept and performance requirements by hiring a design company and give separate orders to companies for detailed design and construction, respectively.

Even though the structure of contract varies, the hierarchy of organisation remains the same – Government (project approver), the project owner, and contractors, including subcontractors. Additionally, each organisation focuses on different levels of work.

In the Korean context, the government has legal authority to approve the master plan of railway projects and accept the railway infrastructure after construction. The Korea National Railway (KNR) is a company established by the Korean government to build national railways and serves as the project owner, as shown in Figure 8. Contractors and subcontractors carry out the detailed design, construction, and stand-alone testing. KNR leads the integration testing of the new railway, and the government will ultimately accept the new railway.

Figure 8 - Hierarchy of organisations for railwway projects

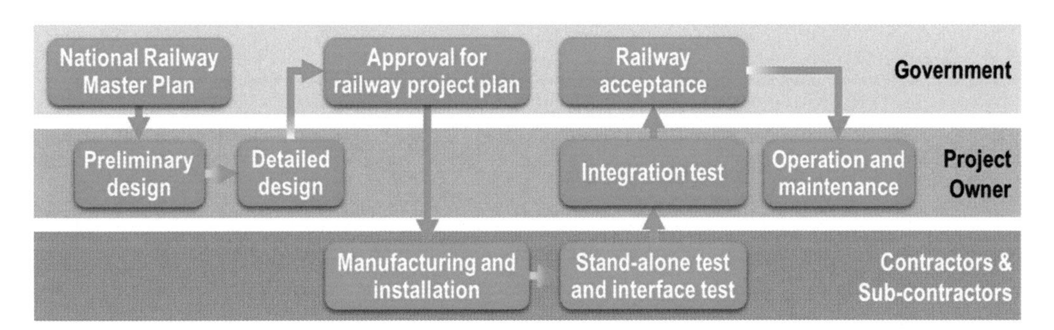

OPERATION PERFORMANCE

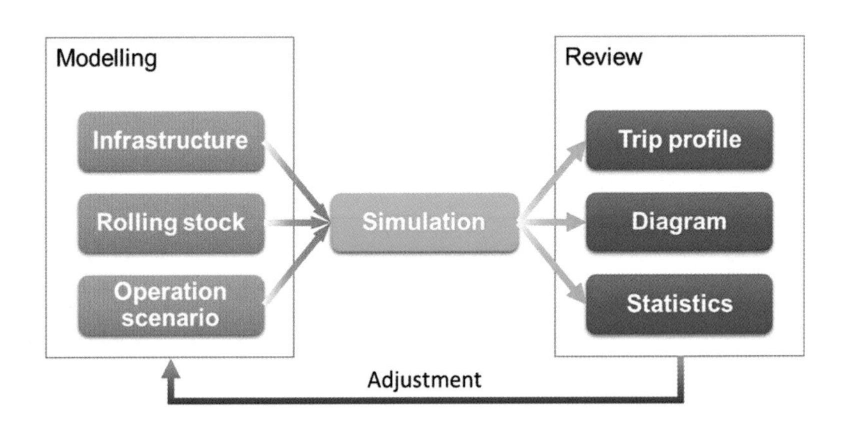

Optimisation of rail service & infrastructure

The cost of building, operating, and maintaining railway infrastructure is significant, so cost optimisation in relation to the level of railway service is a key consideration when developing a railway.

Passenger satisfaction

To continue railway service, profits must be generated through it. From the standpoint of railway owners, the purpose of providing railway service is also to realise profits. These profits come from passengers. Therefore, railway service providers (or railway owners) should consider passenger satisfaction when building a railway. If passengers are satisfied with the railway services, they will willingly choose the railway repeatedly over other transport options.
How can railway service satisfy passengers? The following are the key factors that passengers want from railway service and how satisfaction can be achieved:

- Relatively cheaper ticket price: To lower ticket prices, construction and O&M (Operation & Maintenance) cost should be minimised.
- Rapid transport: This can be achieved by shortening round-trip time (RTT).
- Minimal waiting time: Shortening headway reduces passengers' waiting time at platform. This should be controlled by maintaining minimum headway, especially during peak time.
- No transport service delay (on-time departure and arrival): High service availability can help achieve this.

< Round-Trip Time (RTT) >
RTT is the time it takes for a train to complete the entire route. RTT is estimated as follows:

RTT = Downward trip time + Turnback time at the terminal D +
Upward trip time + Turnback time at the terminal A

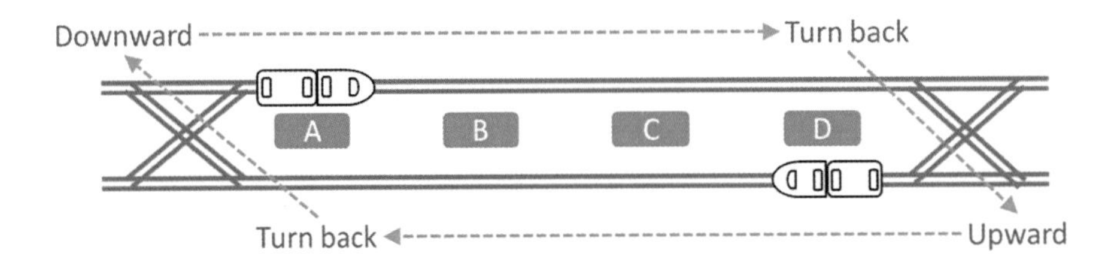

< Headway >

Headway is the interval between consecutive trains. A railway with a shorter headway can transport more passengers than one with a longer headway within a specified time. Headway is directly related to line capacity, which refers to the number of trains that can run on a section within 24 hours. Therefore, the line capacity of a railway with a short headway is greater than that of one with a long headway.

For metro lines, minimum headway is particularly important during peak time, as platforms are generally crowded with commuters during rush hours (peak time) in the morning and at quitting time.

< Service availability >

Service availability in railway service is defined as the probability that the transport service is in a state to perform the required railway service under given conditions at a specific instant or over a specific time interval, assuming that the necessary external resources are provided. As is well known, high service availability leads to high passenger satisfaction.

There are two commonly used types of calculations:

- Reliability-based calculation
- Schedule-based calculation

Both calculation methods will be discussed in the chapter on “RAM (Reliability, Availability, Maintainability).”

< Cheaper price >

The correlation between rail ticket prices and C/O/M (Construction, Operation, and Maintenance) costs is not entirely clear, as most central governments or local governments supress ticket prices. This is because railways and metros are essential public transportation systems. However, rail infrastructure owners must consider the costs of building, operating, and maintaining the railway system to ensure sustainability.

Here is an issue relating to C/O/M costs: when the minimum headway is short, but the round-trip time (RTT) is long, a greater number of rolling stocks is needed for operation. This is because "RTT", "minimum headway", and "the number of rolling stocks" are interrelated by the following equation:

$$\text{Maximum number of rolling stocks} = \frac{\text{Round trip time}}{\text{Minimum headway}}$$

To decrease the number of rolling stocks, RTT must be reduced, or the minimum headway must be increased. If the minimum headway cannot be increased due to passenger satisfaction concerns, the round-trip time must be reduced. Therefore, round-trip time is an important performance indicator that must be managed.

Railway KPIs

Planning personnel in central government or local government will consider how to build an attractive, high-quality railway that requires low construction, operation, and maintenance costs. There are various types of railways that can be built to meet social transportation demands and environmental constraints, as well as numerous methods to optimise different types of railway infrastructure.

Figure 9 - Revenue and cost

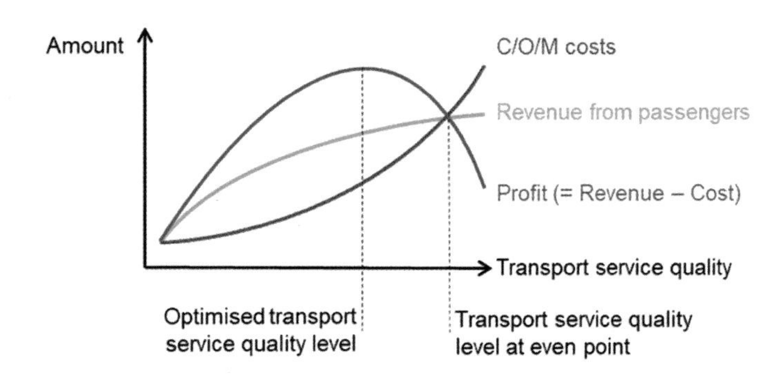

When establishing the operational performance targets, C/O/M costs and revenue from passengers should be considered simultaneously. Increasing operational performance, which can be defined as transport service quality, will lead to higher revenue from passengers; however, C/O/M costs will also increase, as

shown in Figure 9. Regardless of service quality, passengers numbers only increase arithmetically due to capacity limits, while C/O/M costs rise exponentially due to technical difficulties.
Therefore, KPIs relating to operational performance should be carefully studied first and then optimised, as they can support making important decisions more reasonably. Figure 10 outlines how rail KPIs are derived, how they influence one another, and the areas where the systems engineering team can contribute. The arrows indicate the flow of contribution.

Figure 10 - Deriving KPIs

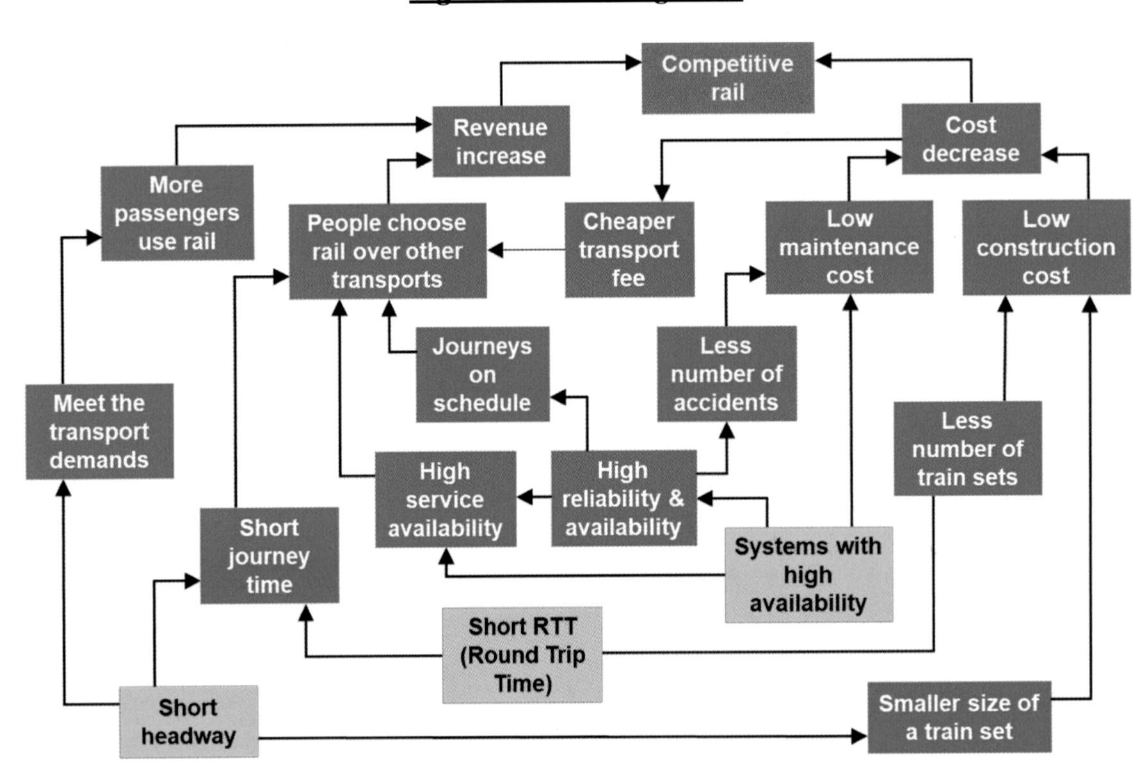

< Revenue increase >

To generate high revenue, people should choose rail over other forms of transport. This can be achieved by providing on-schedule rail service, a short journey time, and high service availability. A short headway can also increase the number of passengers transport.

< Cost decrease >
To save cost, maintenance expenses should be minimised, which can be realised by equipping systems with high availability. Reducing construction costs requires fewer and smaller train sets, which can be realised through a short round-trip time and short headway, respectively.

Minimum headway

Minimum headway refers to the shortest allowable time interval between two consecutive trains traveling on the same track in the same direction. Shorter headways mean increased transport capacity. For example, during peak-time (rush hour), a 10-minute headway allows 6 trains to depart at a station, while 5-minute headway allow 12 trains to depart.

There are two ways to increase number of passengers transported (passenger capacity) during peak time: [1] increasing the number of wagons per train for a fixed number of transport service, or [2] increasing the number of transport service (line capacity) for a fixed number of wagons.

- [1] can be advantageous for saving power and human resources (train drivers) but can be disadvantageous for customer satisfaction due to the longer interval (headway) between trains, which leads to longer passenger waiting time. Few passengers appreciated a long headway, especially during peak times.
- [2] is the opposite of [1] in every way; therefore, metro owners aim to design infrastructure and rolling stock with the shortest possible headway for customer satisfaction.

To shorten headway, engineers consider several design factors in railway planning, including:

- Signalling system – This greatly affects headway, which can be shortened by installing the latest signalling system. For example, 5-aspect signalling systems have shorter headways than 3-aspect systems in fixed-block signalling and CBTC (Communication-Based Train Control) systems allow shorter headways than fixed block systems.
- Performance of rolling stock – breaking distance impacts headway, as the safety distance between trains depends on the breaking distance.

Round trip time

Round trip time refers to the total time taken for a train to travel from its starting point to its destination and back to the starting point. This includes the time spent on the journey in both directions, as well as any scheduled stops or layovers at stations along the route.
Without reducing the line length, round-trip time can be decreased by using higher-performance trains, an advanced signalling system, and well-designed alignment (straightness with minimised inclination). RTT can be predicted by TPS (Train Performance Simulation) and verified through operation tests during the testing & commissioning phase.
RTT can be affected by the infrastructure design factors as follows:

- Line length – The longer the route, the longer the RTT.
- Line straightness – The straighter the route, the shorter the RTT.
- Line gradient – The more slopes the route has, the longer the RTT.
- Number of stations (or stops) – The more stations to stop at, the longer the RTT.
- Dwell time at stations (or stops) – The longer the dwell time, the longer the RTT.
- Signalling system – Although the impact varies depending on the type of signalling system, advanced systems generally lead to a shorter RTT.
- Turn-back time at terminals – The longer the turn-back time, the longer the RTT.

When it comes to the performance of rolling stock, RTT can be affected as follows:

- Maximum speed – The higher the maximum speed, the shorter the RTT.
- Acceleration – The higher the acceleration, the shorter the RTT.
- Deceleration – The higher the deceleration, the shorter the RTT.
- Door opening and closing time – The faster the door opening and closing speed, the shorter the RTT.

On the other hand, planned RTT should be checked during the design phase to determine whether the infrastructure and rolling stock meet the target RTT. Even if it meets the target, it should not be used as a source to develop the operation timetable. This is because dwell time is dependent on passenger congestion at the platform of each station.
Thus, the difference between the two types of RTT should be understood.

- RTT with fixed dwell time at stations – It is planned and checked during the design phase to ensure that the designed rolling stocks can travel the designed railway infrastructure within target RTT. In this case, platform congestion is not a factor, as the RTT includes only the fixed dwell time at each station. This is used to check whether the designed railway infrastructure and rolling stock meet the target.

- RTT considering degree of congestion at platforms – This is used for creating the real operation timetable. If a platform experiences high congestion, the dwell time at the platform should be extended.

Service availability

As addressed in the previous chapter, there are two types of measuring service availability.

< Reliability-based calculation >
The first one is classical reliability-based formula as follows:

$$\text{Service Availability} = \frac{\text{MTBSAF}}{\text{MTBSAF} + \text{MART}}$$

where:

- MTBSAF: Mean Time Between Service Affecting Failures
- MART: Mean Active Repair Time

The processes for calculating the reliability-based formula of service availability will be explained in detail in the chapter "RAM (Reliability, Availability, Maintainability)."

< Schedule-based calculation >
The second one to measure service availability is schedule-based formula, as shown below:

$$\text{Service Availability (\%)} = \frac{\text{Actual trips on time}}{\text{Total planned trips}} \times 100$$

where:
- Service Availability

  - The percentage of the actual trips on time divided by the total planned trips during the pre-defined period.
  - The pre-defined period can be on a daily, weekly or monthly basis

- Actual Trips on Time
    - The number of actual trips arriving at the terminal station on time, not exceeding the pre-defined delay time.
    - The pre-defined delay time is the measured time more than usually 5 minutes excluding mobilisation time.
    - The mobilisation time is the period from the time when a failure is detected to the time before maintenance personnel starts maintenance work.
- Total Planned Trips
    - The number of scheduled trips according to the timetables.

Delays include only those caused by failures of both railway infrastructure and rolling stock, excluding delays caused by overcrowded passengers or any other factors outside the railway system, such as natural disasters. This is because the purpose of measuring service availability is to assess the transportation system's performance purely.

To measure the pure delay caused by the transportation system's failure, mobilisation response time and specific time allowance should be excluded from the total delay time, as follows:

Delay time = Overall downtime – Mobilisation response time – specific time allowance

When a railway system failure occurs, it is not reasonable to regard the entire downtime as delay time because the mobilisation response time to a stalled train or failed subsystem depends on the location of maintenance office – specifically, the distance between the locations. Therefore, the mobilisation response time should be excluded. Specific time allowance refers to the time contracted for a specific reason between the infrastructure owner and the railway development contractor.

When a train fails, all passengers at each station will experience delayed trips at the same time because the failed train will cause delays to other trains as well. In this case, if the line has 14 stations, the delayed trips will be counted as 14 times. This is not reasonable because the number of delayed trips should not depend on the number of stations. From the passengers' standpoint, they have only experienced a delay once. Therefore, it is necessary to find reasonable ways to count delays.
Although the calculation formula will entirely depend on the contract terms and conditions, here are some recommended principles:

- Only the longest delayed departure at a certain station should be counted as a delayed trip.
- As soon as a train breaks down, passengers at a station will experience the longest delay time of departure. This delay time will gradually reduce as recovery driving takes place. In this case, any delayed departure exceeding the allowed time (generally 5 minutes) should be counted as a delayed trip.

Framework of railway service

Preliminary operation and maintenance plan

At the beginning of Systems Engineering work, it is necessary to establish the concept of railway operation and maintenance, which will be considered when working on RAM, safety, requirements, interface, etc. The preliminary operation and maintenance plan should include the service target. Based on those preliminary plans, each manager (safety, requirements, etc.) of systems engineering will prepare a management plan, and the outcomes of each plan can provide feedback to update the preliminary plans. The updated preliminary plans will then serve as the input for preparing detail operation and maintenance plans.

< Preliminary operation plan >

The preliminary operation plan will include the following contents:

- Operation organisation
- Passenger service characteristics: operating hours, operation headway, station dwell time, round trip time, train size (including spare vehicles), etc.
- System operation modes: normal operation modes, degraded operation mode, etc.
- Operation plan: normal operation, degraded operation, emergency operation, etc.
- Failure management: detection control function in OCC, manual operation, delay time in self-propelled failure train, recovery of stalled trains, train stop, etc.
- System degraded operation: power supply and distribution system, etc.
- System start-up and shutdown

< Preliminary maintenance plan >

The preliminary maintenance plan will have the following contents:

- Maintenance organisation: divisions, shift patterns, etc.
- Maintenance guidelines: rolling stock, track work, power supply, signalling, communication, rolling stock maintenance equipment, etc.
- O&M tools and equipment
- Spare parts and consumables: rolling stock maintenance equipment, trackwork, power supply, signalling, communication, stations, depot, etc.

- Maintenance management system (MMS): major functions, key performance indicators, MMS outputs reports, annual maintenance plan, etc.

Operation modes

Operation modes can be divided into three: normal operation mode, degraded operation mode, and emergency mode. The purpose of classing operation modes is to establish strategies for each mode.

< Normal operation mode >
Normal operation mode is the operation being conducted as scheduled or according to timetable. Scheduled speed, minimum headway, round-trip time, service availability, and other KPIs are established considering the normal operation mode. Train operation simulation is also performed for this mode.

< Degraded operation mode >
Degraded mode refers to a situation where normal transportation service is not possible due to certain failures of railway system, such as signalling, power supply, or rolling stock. It differs from emergency mode: evacuation is the top priority in an emergency, while continuing service with reduced performance is the top priority when part of the railway system fails. For degraded operation mode, multiple operation scenarios should be prepared to allow bypassing failed areas via crossovers if necessary.
Operation scenarios for degraded operation mode can be produced as follows:

- Design review – whole framework of the systems of the line is reviewed.
- FMECA development – failure, effect, and criticality are analysed.
- Critical subsystems/areas definition – critical subsystems and areas are defined based on FMECA results.
- Scenario development – scenarios are developed to conduct a bypass or a detour.
- Performance analysis – performance is analysed based on the degraded operation scenarios.

< Emergency mode >
Emergency mode arises not from the system failures but from external causes such as disasters, terrorism, fire, etc. Since emergency is an uncontrollable situation, two types of strategies should be prepared: a passenger evacuation plan and a response system for crisis situations. Since emergency cases can vary significantly, it is not possible to establish a single system lockdown procedure and apply it universally. An evacuation procedure will include:

- From rail vehicle to safe zone: Passengers in a vehicle will be guided by a vehicle driver, announcement from OCC, and/or operation personnel until they arrive at the safe zone.
- From station to safe zone: Passengers at a station will be guided by announcement and/or station staff until they arrive at the safe zone.
- Measures for transfer of casualties: Some cases require measures for the transfer of casualties.

Scheduled speed

Scheduled speed is the ratio of the distance between two specific stations (typically terminal stations) to the total running time, including the dwelling time at the platform. In contrast, average speed does not account for the dwelling time. Scheduled speed is closely related to Round-Trip Time (RTT) and serves as a basis of operation planning for the normal operation mode.

$$\text{Scheduled speed} = \frac{\text{Distance between terminal stations}}{\text{Running time} + \text{Dwelling time}}$$

Scheduled speed is influenced by the design of infrastructure and the performance of rolling stock, like RTT, apart from line length and turn-back time at terminals. Therefore, if the target scheduled speed (or RTT) cannot be achieved, the following factors should be improved: line straightness, line gradient, number of stations (or stops), and signalling system. Please refer to the chapter “Round trip time.”

< Design speed >

Many people are confused with the terms ‘scheduled speed’ and ‘design speed.’ Design speed in railway projects refers to the speed at which railway infrastructure should be designed, considering train axle load. Since it relates to structural safety, the design speed is typically about 10% greater than scheduled speed.

Headway, round trip time, and train sets

Before addressing operational matters, we need to understand how minimum headway, round-trip time, and the quantity of train sets are determined. These are all calculated based on expected transport demand. This chapter explains how they are calculated.

Approaches to railway design

When planning to build a railway, the first step is to analyse the demands of railway service, which is generally conducted through a feasibility study.

In the railway planning phase, there are two factors to be balanced to build an optimised railway as follows:

- Minimum headway: it affects the required passenger capacity of a train set and the number of train sets for operation.
- Passenger capacity per train set: it affects the number of operations per hour during peak-times.

Both factors mentioned above can be traded off, and planning engineers should optimise the railway service plan by balancing them. Sometimes, one of them is fixed first due to various project environmental conditions. If the passenger capacity of a train set is fixed, the summarised process to calculate the quantity of train sets is as follows (Please note that this process is a summary!):

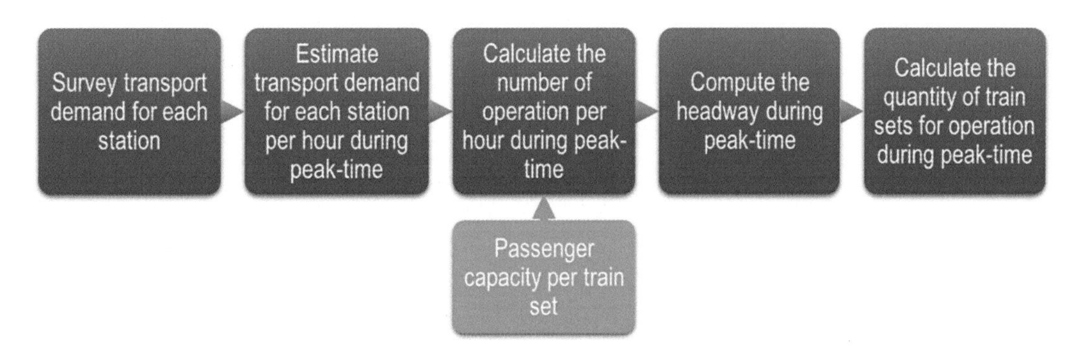

On the other hand, when headway is fixed, the summarised process for calculating the passenger capacity per train set is as follows (Please note that this is a summary!):

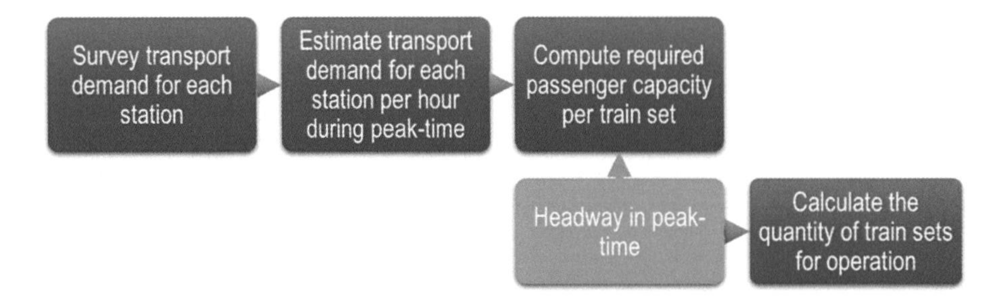

Detailed process for estimating demands

Regardless of whether the headway is fixed, or passenger capacity of a train set is fixed, the detailed process for defining the target transport demand is shown in Figure 11:

Figure 11 - Process for defining the target transport demands

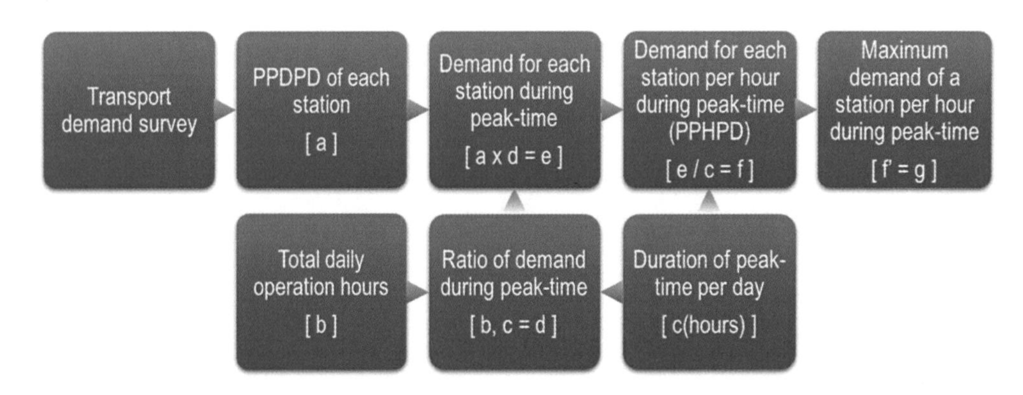

Combining the transport demand survey with other operation factors, such as daily operating hours and peak-time duration, allows for the estimation of the number of passengers per day per direction (PPDPD) at each station. PPDPD represents the total number of passengers expected to use train services at a particular station for a specific direction daily.

Figure 12 - An example of PPDPD

Station (A⇒H)	Boarding	Alighting	in Train
A	6,000	-	6,000
B	14,500	3,300	17,200
C	27,200	15,200	29,200
D	28,400	28,600	29,000
E	18,200	26,100	21,100
F	9,300	17,500	12,900
G	1,200	7,900	6,200
H	-	6,200	-

Station (H⇒A)	Boarding	Alighting	in Train
A	-	7,000	-
B	1,500	7,700	7,000
C	9,500	16,500	13,200
D	19,200	28,100	20,200
E	27,400	26,600	29,100
F	25,200	14,200	28,300
G	14,400	3,200	17,300
H	6,100	-	6,100

Based on the total daily operating hours and peak-time duration, the ratio of demand during peak-time should be estimated. This ratio indicates the percentage of the total number of passengers predicted to use train services during rush-hour. For instance, if a particular station has 100,000 passengers per day and the peak-time passenger ratio is set at 40%, the expected number of passengers using the station during peak-time would be 40,000.

The ratio of the peak-time demand for a metro line is relatively higher than that for an intercity line. Since metro lines are primarily used for commuting, they naturally experience greater passenger demand during rush-hours compared to intercity lines. However, this ratio can vary and is dependent on local conditions, such as the number of commuters.

Considering the rail service type, the peak-time ratio should be defined, and then the demand at each station during peak time is calculated, as shown in Figure 13. In the example, the applied ratio is 60%.

Figure 13 – An example demand during peak-time

Station	A⇒H	H⇒A
A	3,600	-
B	10,320	4,200
C	17,520	7,920
D	17,400	12,120
E	12,660	17,460
F	7,740	16,980
G	3,720	10,380
H	-	3,660

To help readers understand the process easily, the operation plan of an example railway project is defined as follows:

- The line is a metro service providing commuter transport.
- Operations begin at 05:00 and finish at 24:00, resulting in a total daily operation time of 19 hours.
- The duration of peak-time is from 07:00 to 09:30 in the morning and from 17:00 to 19:30 in the evening, amounting to a total daily peak-time duration of 5 hours.

PPDPD should be converted into PPHPD (Passengers Per Hour Per Direction), as PPHPD is used for estimating various factors such as passenger capacity, headway, etc. By dividing PPDPD by the total daily peak-time (5 hours), the PPDPD for each station can be obtained, as shown in Figure 14.

Figure 14 - An example of PPHPD at each station

Station	A⇒H	H⇒A
A	720	-
B	2,064	840
C	3,504	1,584
D	3,480	2,424
E	2,532	3,492
F	1,548	3,396
G	744	2,076
H	-	732
Max	3,504	3,492
Target	3,504	

The maximum demand among the stations should be defined, which then becomes the target to be met. In Figure 14, the target demand is 3,504 for station C in A⇒H direction. Based on PPHPD, the size of each station and the volume of facilities and equipment at each station are designed accordingly.

To satisfy the target transport demands, engineers must recognise two key factors typically considered when planning a railway:

- Line capacity – the number of trains that can operate on a section within 24 hours, which primarily depends on the minimum headway.
- Passenger capacity – the number of passengers that can be accommodated by a train set.

Line capacity, passenger capacity per train set, and transportation demands (PPHPD) are interrelated as follows:

PPHPD = Line capacity per hour per direction × Number of passengers per a train set

For the scenario depicted in Figure 14, if the line permits trains to pass a section up to 10 times per hour, each train must have the capacity to transport more than 351 passengers, calculated as follows:

10 times x 351 passengers = 3,510 pphpd [3,510 > 3,504]

If each train can transport up to 710 passengers, the line capacity would only require 5 trips to meet the demand. The calculation is as follows:

5 times x 710 passengers = 3,550 pphpd [3,550 > 3,504]

This demonstrates that both line capacity and passenger capacity can be traded off to optimise construction cost and O&M (Operation and Maintenance) costs.

To optimise the railway operation target, the headway and passenger capacity per train set should be balanced.

- If the line capacity is increased, the headway will be decreased which may consequently satisfy passengers since everyone prefers a shorter waiting time. However, this will lead to an increase in the number of train sets required, resulting in a significantly higher budget.
- Conversely, if the passenger capacity per train set is increased, the headway can be extended. This will have the opposite effect of the previous case, allowing for reduced number of train sets and potentially lowering overall costs.

Therefore, both factors should be balanced simultaneously to optimise the infrastructure against the transport demand. However, the range for balancing these factors is limited for several reasons:

- Headway during peak-time: it is challenging to reduce the headway to less than 4 minutes due to technical limitations, while extending it beyond 10 minutes is generally unreasonable for metro lines. The acceptable range for headway varies depending on the railway service type and commuting environment.
- Passenger capacity per train set: The options for changing the formation of a train set are limited because the number of wagons forming a train set must be an integer. For instance, if calculations suggest a requirement of 5.6 wagons per train set, the configuration must be rounded to 6 wagons, as fractional wagons are not permissible.

According to project conditions, one is fixed first, and then the other is calculated as follows:

Case – when the passenger capacity per train set is fixed

When the passenger capacity per train set is fixed first, the following values can be calculated as outlined in the process:

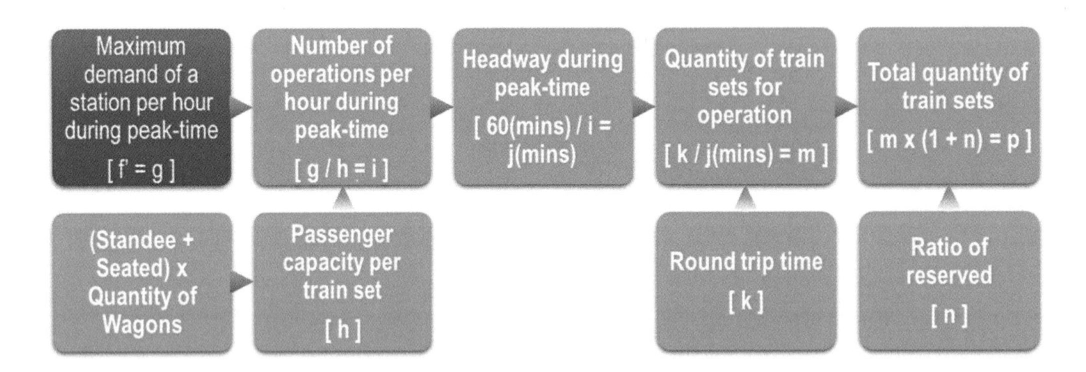

< Number of operations per hour during peak-time >

- If the railway service can handle the maximum transport demand of a certain station, it will also be able to manage the demands of the other stations as well.
- The required operations can be estimated by dividing the maximum hourly peak-time demand of a station by the passenger capacity per train set. For example, if a train set can carry 477 passengers, the required operations would be 7.3 trips.

< Headway during peak-time >

- Since headway (in minutes) directly affects passengers' waiting time on the platform, planners should optimise headway by balancing construction costs and passenger satisfaction.
- Headway is calculated by dividing 60 minutes by the number of operations per hour during peak-time. For instance, 60 minutes divided by 7.3 trips equals 8.2 minutes. For simplicity in operation planning, 8.0 minutes will be used.

< Quantity of train sets for operation >

- The maximum number of train sets operating simultaneously during peak-time should be determined.
- It is calculated by dividing the round-trip time (in minutes) by the headway (in minutes) during peak-time. For example, if the headway is 8 minutes and the round-trip time is 120 minutes, 15 train sets are required.
- Train operation simulation tools are used to calculate the round-trip time.

< Total quantity of train sets >

- The total number of train sets should include the number of reserved ones to prepare for rolling stock failures and heavy maintenance.
- It is calculated by multiplying the quantity of train sets for operation by the sum of the ratio of reserved and 1. For instance, if 15 train sets are required for actual operation and the reserved train set ratio is 10%, a total of 17 train sets will be required. Even though the actual number is 16.5 (= 15 x 110%), the required should be an integer.

Case – when the headway during peak-time is fixed

When the headway during peak-time is fixed first, the following values can be calculated as outlined in the process:

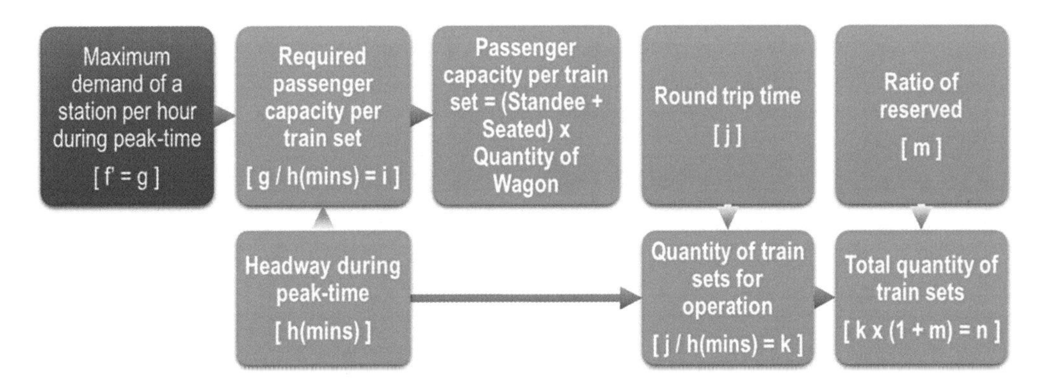

< Required passenger capacity per train set >

- It is estimated by dividing the maximum demand of a station per hour during peak-time by the headway during peak-time (minutes). For instance, if the target demand is 3,504 and the headway is 5.0 minutes, the required passenger capacity per train set is 700.8 (= 3,504 / 5).
- This value is used to define the formation of a train set (See Figure 15). In the example, a train configuration consists of 150 standees and 50 seats per wagon. By adding wagons, the passenger capacity can be increased. After performing the calculation, 3-wagon and 4-wagon train types are viable. Since the target capacity is 700.8, 3-wagon train type will be selected.

Figure 15 - An example of train compositions

Standee	Seated	Qty of wagon	Total	150%	Decision
150	**50**	**1**	**200**	**300**	**not**
150	**50**	**2**	**400**	**600**	**not**
150	**50**	**3**	**600**	**900**	**accepted**
150	**50**	**4**	**800**	**1200**	**accepted**

Other factors, such as the quantity of train sets for operation and the total quantity of train sets, are calculated in the same manner as in the case where “the passenger capacity per train set is fixed”.

In conclusion, operational performance estimation encompasses headway, round-trip time, and PPHPD. Operation performance management involves overseeing these three factors.

Train operation simulation

In the design phase, train operation simulation should be conducted to verify whether the planned railway infrastructure can achieve the operational targets of round-trip time and headway. As highlighted in the previous chapter, round-trip time is crucial because it determines the number of train sets required, which directly impacts on the project costs.

Simulation process

There are various software tools available, such as OpenTrack, that offer operation simulation functions. These tools are typically used to:

- Estimate the round-trip time
- Identify bottlenecks with minimal construction efforts
- Assess the line capacity of a railway line
- Create timetables
- Evaluate the impact of changes on railway infrastructure

To simulate the train operation, the following steps should be performed (See Figure 16 for the overall process):

- Model the line (infrastructure) layout and alignment.
- Configure the performance value of rolling stock.
- Set up the timetable.
- Perform the simulation.
- Review the output data.

Figure 16 - Operation simulation process

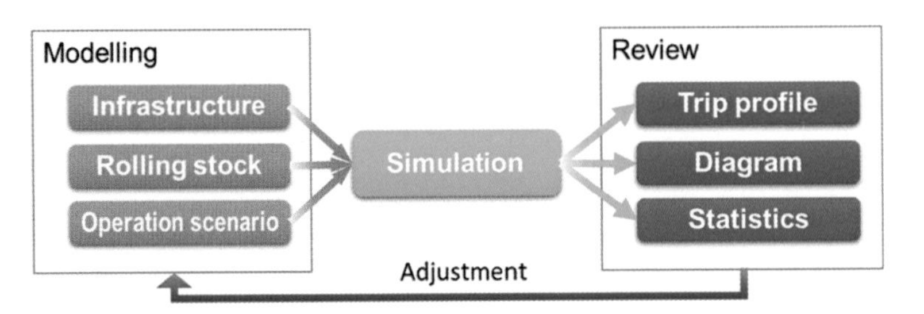

To model a line, the following data should be gathered or determined at least:

- Track: length of the line, alignment (curvature, horizontal transition, gradients, and vertical transition), location of stations, switching & crossings, and speed limits.
- Signalling: system type (aspects), location of signals, location of track circuits, and axle counters.
- Rolling stock: train type, length of a train set, axle load, traction power, maximum speed, acceleration, and deceleration.
- Operation scenario: dwell time at each station, turn-back scenarios at terminal stations, headway, and driving performance.

After performing the simulation, the software can produce the following data:

- Speed profile – demonstrates speed at each point of the line.
- Diagram – like a trip diagram.
- Track occupancy – checks for conflicts of allocations.
- Statistics – includes metrics such as traction power at each point of the line.

Railway designers will review the data produced by the software. If they find the results do not satisfy operational requirements, they can adjust parameters in the model and conduct the simulation repeatedly until the results meet the operational criteria.

Modelling and simulating by using OpenTrack

This chapter describes how to model a railway infrastructure and simulate it by using OpenTrack.

< Creating Infrastructure >

(1) Create vertices that contain data about kilometre points and names.
(2) Connect the vertices with edges that contain data about gradients, radii, speeds, and track names.
(3) Place signals defining their attributes – block signals, home signals, exit signals, distant signals, and stop centre.
(4) [User will draw a long railway line separating it into several segment for easy viewing and this step is for how to connect them] Connect the separated line using the "Connector".
(5) Create station database and place a station by selecting a station from the database.
(6) Define a station area, within which all vertices will turn light blue once the area is established.
(7) Define routes that are automatically connected from a main signal to another.
(8) Define paths that are automatically connected from an exit signal to the next or from a home signal to another.
(9) Create itineraries from the first station to the last station (terminal stations).
(10) Generate courses and services by selecting itineraries.
(11) Create a new timetable.
(12) Group courses if needed.

< Creating rolling stock >

(1) Generate a new train set
(2) Input train data, such as the length of the train set, maximum speed, weight of the train set, maximum acceleration, maximum deceleration, etc.

< Running a train – simulation >

(1) Create the corridors.
(2) Start the simulation.
(3) Plot the simulation results by selecting the data you want to check.

Review of simulation results

After the simulation, the results can be viewed in various formats, including

- Distance-speed diagram
- Time-distance stepping diagram
- Power consumption diagram,
- Other relevant output data

< Distance-Speed Diagram >

In the distance-speed diagram, the horizontal axis represents distance, and the vertical axis indicates speed. The grey line denotes the speed limit, and the red line represents the travelling speed. This diagram illustrates how a rolling stock travels along the alignment, highlighting the sections where it can accelerate or decelerate.

Figure 17 - An exmaple of Distance-Speed Diagram

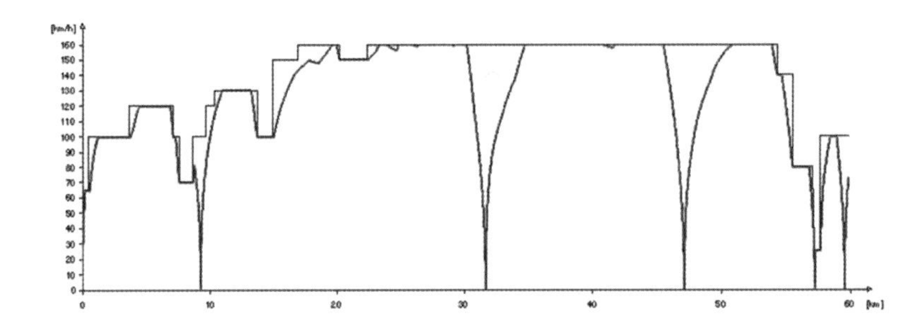

< Time-Distance Stepping Diagram >

In the distance-time stepping diagram, the horizontal axis represents time, while the vertical axis indicates distance. The grey lines represent the headways of the preceding and following trains, while the red lines indicate the locations of the trains at specific times.

Figure 18 - An example of Distance-Time Stepping Diagram

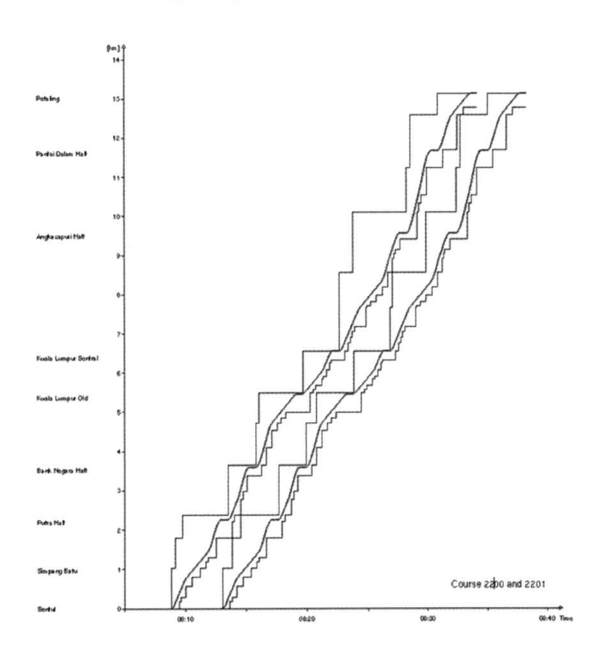

< Power Consumption Diagram >

In power consumption diagram, the horizontal axis represents distance, while the vertical axis indicates the capacity of the power supply. The red line shows the required power capacity, while the blue line represents the planned capacity of the power supply for each section.

Figure 19 - An example of Power Consumption Diagram

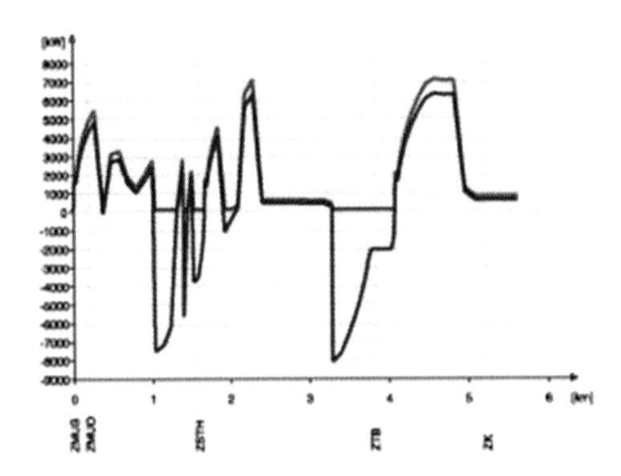

Reviewing the simulation results may lead to design changes aimed at meeting the operational requirements or optimising the railway design. To shorten the Round-Trip Time (RTT), potential design changes may include the following:

- Track: Minimise the length of the curved sections, enlarge the curvature, minimise the length of the slope sections, reduce the slope angle, and decrease the turn-back time at the terminals.
- Rolling stock: Increase acceleration, maximum speed, and deceleration, while reducing the door opening and closing time.

REQUIREMENTS MANAGEMENT

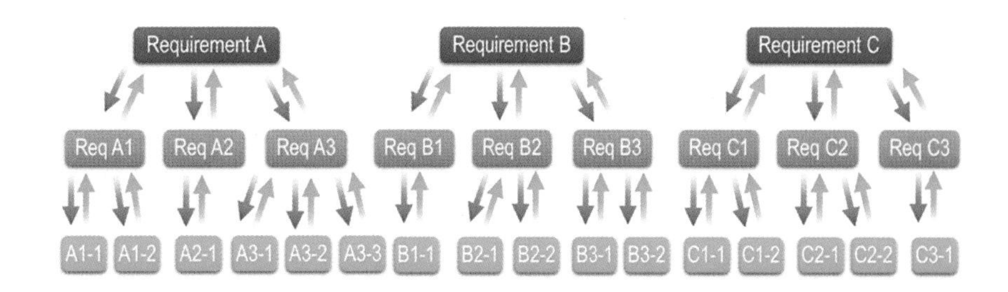

Overview

Requirements in Systems Engineering are statements that define characteristics of a system, product, or service to satisfy contractual obligations, laws & regulations, technical specifications, and environmental conditions. Contractors or developers must verify and validate these requirements throughout the project life cycle contractually and technically.

Why requirements management process?

In the Testing and Commissioning (T&C) phase, non-compliance items can occur due to:

- Engineers misunderstanding the project requirements.
- Errors in the technical aspects themselves.
- Interface and/or interference errors.
- Neglecting certain requirements for various reasons.

According to requirements management specialists, 40~50% of non-compliance stems from engineers misunderstanding the requirements. However, engineers can mitigate their misunderstanding and neglecting requirements by following a structured requirements management process. While performing the requirements management process, the scope of the project becomes clear, and issues like interface and safety are easily derived. Requirements management is related to scope management and interface management, including other requirements such as safety, human factors integration, EMC (Electro-Magnetic Compatibility), etc.,

Framework of requirements management

The framework for requirements management primarily consists of requirements database, requirement items, and control methodologies, as illustrated in Figure 20.

Figure 20 - Framework of requirements management

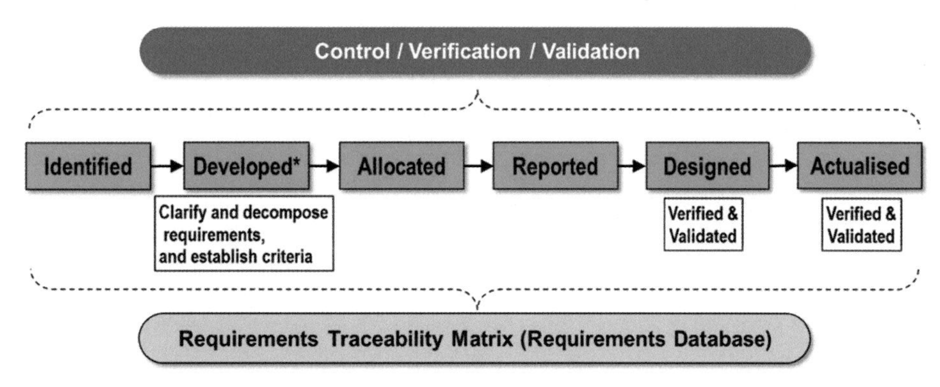

To manage the traceability of requirements, establishing a database is essential due to the typically large volume of requirement-related information. All requirement-related data should be stored in a dedicated database. However, this does not imply that every requirement manager must use specific software; for smaller volumes of information, Microsoft Excel may serve as a more effective management tool than specialised databases. All requirements will progress through the stages of identification, development, allocation, reporting, design, and actualisation. Throughout these stages, the requirements will be controlled, verified, and validated with the output of the requirements management activities also stored in the database.

Types of requirements

Requirements can be classified into the following types:

- Contractual requirements: These include requests for proposal (RFP), contractor proposal, and contractual documents.
- Mandatory requirements: These are imposed if specified in the contractual documents, including laws and the client company's internal regulations.
- Technical requirements: These are applicable if included in the contractual documents, encompassing technical specifications, domestic standards, and international standards.
- Environmental requirements: These are also contingent upon the contractual documents, covering natural environmental conditions and social conditions.
- Outcomes from the previous phase: These may include preliminary design.

Figure 21 shows the contents of both technical and non-technical requirements. Since non-technical requirements, such as the project schedule, can sometimes affect the quality of technical ones, they should also be managed.

Figure 21 – Technical & non-technical requirementss

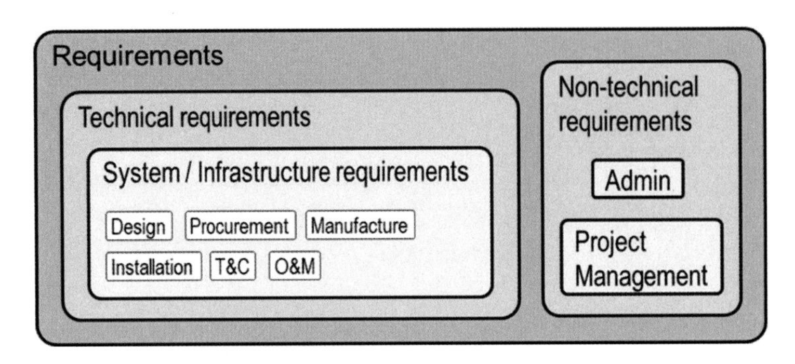

Requirements management process and structure

Below is the requirements process:

- Gather and review all contractual documents.
- Define and clarify all requirements.
- Decompose each requirement (in a downwards direction as shown in Figure 22) until a sub-requirement can be designed and actualised by sub-contractors and suppliers. Thus, a sub-requirement at the lowest level should be verifiable.
- After design and manufacture, each deliverable should be verified and/or validated with evidence (in an upwards direction as shown in Figure 22). At the upper level, requirements are finally checked with verification/validation evidence.

Figure 22 - Requirements decomposing and integration integrating

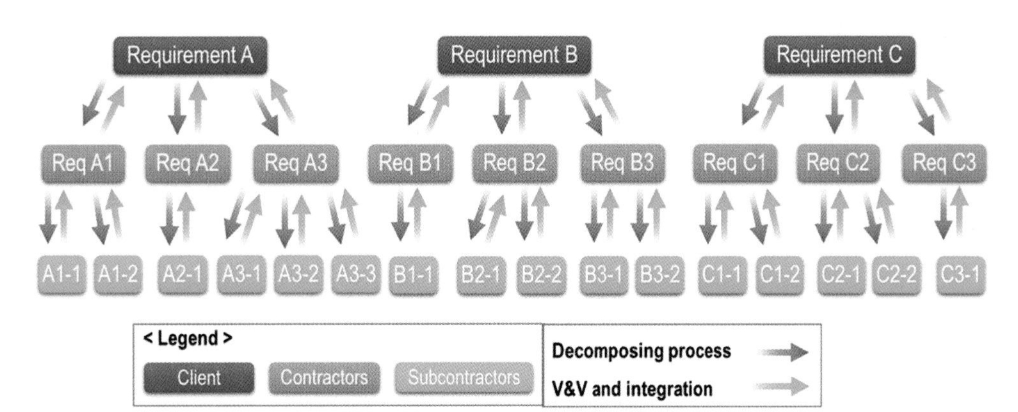

From the standpoint of the system integrator at the top level in the organisational structure of the railway projects, it is not reasonable to manage all requirements from the top-level requirements governed by client to the lowest level requirements controlled by sub-contractors and suppliers. This approach is inefficient. In the case of the system integrator hired by the client, their scope of work regarding requirement management will cover only the requirements allocated to the contractors hired by the client, not those allocated to the subcontractors hired by the contractors. As shown in Figure 22, the system integrator can manage the A and A1 levels. However, if A1-1 is transferred from a contractor to its subcontractor, the responsibility for control shifts to the contractor. The depth of the system integrator's responsibility will be determined based on contractual relationship.

ICE and IV&V(E)

Clients in huge railway projects often engage Independent Checking Engineers (ICEs) or Independent Verification & Validation Engineers (IVVEs) to independently assess contractors' deliverables in terms of requirements satisfaction and product quality.

< ICE >

ICEs typically provide verification services that include the following:

- Compliance of the contractor's scope of work with the client's requirements.
- Compliance of the contractor's design with the applicable contractual standards.

- Correctness and completeness of the contractor's design deliverables.
- Compliance of the contractor's construction techniques and procedures with the applicable standards.
- Correctness and completeness of the contractor's construction deliverables.

< IV&V (Independent Verification & Validation) >

IV&V refers to verification and validation activities conducted by a team or individual independent of the contractor leading the project. Since IV&V team must review and confirm that the requirements are correctly defined and fully satisfied, they should be technically, managerially, and financially independent of the contractor(s).

All aspects of the contract for a specific project are associated with requirement-related issues from a requirements management perspective. Therefore, IV&V has been widely adopted to efficiently manage contractors' deliverables in construction projects in recent years.

According to EN 50129, high levels of SIL (Safety Integrity Level), such as SIL 4, necessitate the involvement of independent verifiers and validators who are not under the control of the project manager. Given the requirements of EN 50129, the presence of IV&V becomes particularly essential to ensure that a product or a system rigorously meets safety and quality standards.

The roles and responsibilities of ICEs and IVVEs appear to be similar. However, while ICEs focus on technical verification activities related to the contractor's design and construction deliverables, IV&V encompasses a broader scope, covering all contractual requirements, including both technical requirements and non-technical aspects, such as schedules (timescale).

Requirements identification and clarification

The first step in requirements management is to identify and define requirements from contractual documents, relevant laws, regulations, standards, and specifications. All identified requirements must be clarified.

Requirements identification & clarification

< Identification >

The scope of work is outlined in the contractual documents, so the requirements manager should review these documents first. As introduced in the previous chapter, requirements within the contractual documents can be classified into technical requirements and non-technical categories. Non-technical requirements pertain to administrative aspects, while technical requirements must be reflected in the system development. Technical requirements typically encompass functionality, quality, and safety.
During the requirements identification stage, it is essential to review relevant standards and regulations mentioned in the contractual documents. Additionally, several types of requirements should be considered in the design stage, even if they are not explicitly stated in the contractual documents, including:

- Procedural requirements: laws, client's organisational regulations, etc.
- Technical requirements: domestic standards, international standards, etc.
- Environmental requirements: natural environmental conditions, social conditions, etc.

Once all the target requirements have been collected, the statements should be divided into sentences. These sentences should then be categorised as either actual requirements or information. Requirement-related documents often contain both. Requirements are the actions the contractor must take, while information is what the contractor simply needs to reference. One of the easiest ways to distinguish them is by identifying sentences that contain words like "shall", "should", or "must," which typically indicate requirements. However, sentences without these words could still contain requirements, as references can sometimes also be requirements.

< Clarification >

Even if all requirements are identified from the relevant documents, it may not always be possible to act on them immediately, as some sentences may be vague and open to interpretation. Such ambiguities

can lead to disputes between the client and contractor. Engineers at higher levels in the organisational structure often face these challenges. A requirements manager should review the requirements, identify vague sentences, and work with engineers to clarify them.
The ambiguity of a sentence in a technical work can arise from two types of sentences as follows:

- A sentence that can be interpreted in many ways. These vague sentences should be re-written so that both the client and the contractor have a shared understanding of the requirements.
- A sentence (requirements) that suggests a broad-range requirement. These requirements should be broken down into smaller, more specific, and measurable requirements. A good example for such requirements is something like that "the machine shall work in a proper manner." However, there are no criteria to define "a proper manner." To make the requirement measurable and verifiable, it is necessary to decompose it into several sub-requirements such as availability, electromagnetic interference with other machines, hazards aspect, etc.

If possible, the re-written sentences should be approved (or at least reviewed) by the client, as the interpretation of requirements significantly impacts the works. The requirements manager should maintain both the original sentences and re-written sentences in the requirements database.

< Criteria >
Once all sentences in the target requirements are clarified, acceptance criteria for each requirement should be established to make the requirements verifiable. These criteria will serve as the basis for verifying each requirement. The acceptance criteria should also be clear to avoid disputes between the client and the contractor. The most crucial characteristic of both the requirements and criteria is that they must be measurable or verifiable, as they need to be verified and validated at the end of each phase.

Premises & assumptions

Premises and assumptions (P&A) refer to considerations that should be defined at a certain point but have not yet been defined due to the project's lack of maturity. Since these P&A can significantly impact requirements management, they must be made as clear as possible and, if feasible, officially documented.

When plans are being produced at the beginning of a project, there can be information (or considerations) to be reflected in the plans but that are not yet well defined. In such cases, those responsible for developing the plans may hesitate due to the lack of defined information. To address this, they need to prepare P&A

documents describing the undefined matters. Project managers, system integrators or risk managers should then collect all premises & assumptions from the plans and formalise them, as P&A represent hidden risks.

<u>Figure 23 - Descrease of premises & assumptions</u>

Concept	Preliminary Design	Detailed Design	Installation	T&C	O&M

Defined requirements

Premises & Assumptions

As the project progresses, the number of P&A will ideally decrease, and there will ideally be no P&A by the end of detailed design, as shown in Figure 23.

Requirements development and apportionment

Some of requirement statements may not be specific enough to be applied to design, procurement, installation, etc. Additionally, some requirements may need to be considered by some subcontractors simultaneously when designing. Those types of requirements should be decomposed and allocated to each design team.

Principles of requirements development

The purpose of requirements development is to make each requirement applicable and verifiable. The requirements development principles are as follows:

- Divide a simple and broad requirement into several specific ones: A requirement like "Install a safe elevator," for example, is too simple to apply to design. In this case, engineers should divide it into several applicable ones, such as "The contractor shall procure and install a safety-certified elevator" and "The elevator shall have the function of stopping operation in the event of a failure."
- Change non-measurable requirements into measurable ones: Any requirements with numbers are easy to measure.
- Make a time-bound requirements: It is recommended to indicate the phase in which the requirement should be actualised or met, which data should be stored in the requirements database.
- Ideal requirements are SMART (Specific, Measurable, Achievable, Relevant, and Time-bounded).

If each requirement has become achievable and verifiable, they all can be called the requirements specification. The requirements specification can consist of various types of documents, such as requirements statements, calculation sheets, drawings, spreadsheets, requirements database, etc. The important thing to remember in this process is that the requirements specifications developed by engineers must be approved by the client. If the client does not want to take the process to approve the requirement specifications, it must at least be reported to the client. This is because all contractors' activities will be based on the specifications.

Requirements decomposing and allocating

No matter whether a contractor has sub-contractors or not, requirements should be decomposed. However, not all requirements should be decomposed. Figure 24 outlines the management level of requirements;

blue coloured requirements are drawn from the contractual documents between the client and the contractor; grey ones are developed by the contractor, and sky-blue ones by the sub-contractor(s). Most importantly, the requirements at the lowest level (sky-blue ones) must be measurable for the purpose of V&V process. The sub-requirements at the lowest level should have criteria for verification and validation.

Figure 24 - Requirements decomposing

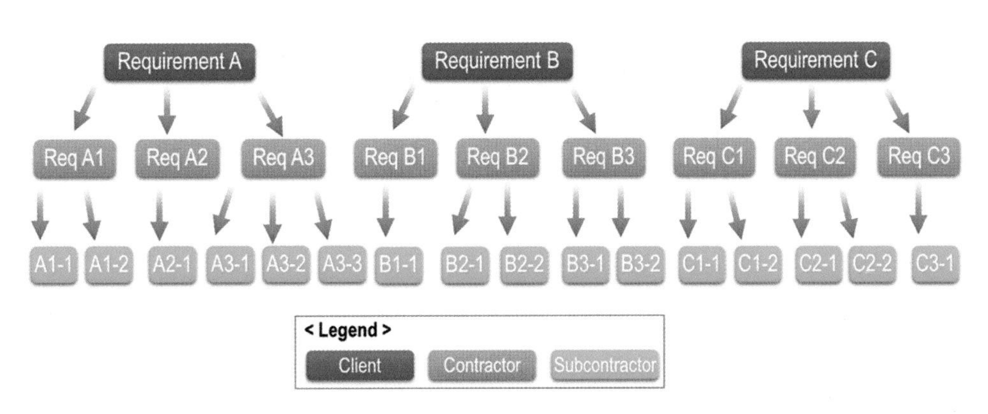

Requirements allocation is one of the hardest processes because each team is willing to accept the least amount of responsibility for the requirements. If a team does not fully accept the assigned requirements and only selectively accept some of the requirements, the requirements manager needs to ask sub-contractors to justify the reasons for the unaccepted requirements assigned to the team.

< Criteria decomposition >

Criteria for each sub-requirement must be defined by decomposing the upper-level criteria in advance to be used in V&V phase. Since a verifier and validator will conduct V&V, criteria prepared by subcontractors or suppliers should be confirmed for its suitability by the verifier and validator.

To ensure that requirements are verified and validated effectively, the requirements criteria must be clear, measurable, and testable. This principle emphasizes that requirements should be:

- Clear: Unambiguous and easily understood by all stakeholders.
- Measurable: Quantifiable or evaluable through specific metrics or tests.
- Testable: Capable of being tested to confirm that the requirement has been met.

This helps to facilitate accurate verification (which confirms that the system meets the specified requirements) and validation (which ensures the system fulfils its intended use).

Process for Key Performance Indicators

A Key Performance Indicator (KPI) is a measurable value that demonstrates how a goal is achieved. Railway clients or project owners define KPIs to control their railway projects, as outlined by the following examples (Please refer to earlier chapters): Round-Trip Time, Minimum Headway, Service Availability, Systems Availability, Fatalities, etc.

Service availability and systems availability should be predicted during the development phase and demonstrated in the operation phase after opening through the conduct of FRACAS. They will be introduced in the chapter "RAM (Reliability, Availability, Maintainability)".

< Process >

In requirements management process, KPIs should be managed and monitored using different approaches because they are managed not just by controlling specific components but by controlling the entire system. The management process for KPIs may include the following steps:

- Define a formula for each KPI.
- Decompose factors of a KPI formula and allocate each factor to the respective subcontractor (to each subcontractor's deliverables, specifically).
- Measure the performance of each subcontractor's deliverable.
- Check whether the overall value satisfies the target KPI.

If a target KPI has multiple factors like:

$$\text{A target KPI} = \text{Factor}_1 + \text{Factor}_2 + \text{Factor}_3 + \cdots + \text{Factor}_n$$

or

$$\text{A target KPI} = \text{Factor}_1 \times \text{Factor}_2 \times \text{Factor}_3 \times \cdots \times \text{Factor}_n$$

or mixed

Each factor is assigned to each subcontractor as their target KPI to achieve. When allocating the factors, however, the contractor must consider the technical constraints.

Here is a good example:

When a contractor has a target service reliability of 98.8% monthly, it means that the contractor is allowed to have just one failure per month. If the contractor has 5 subcontractors – A, B, C, D, and E – that all evenly affect the reliability of the entire system, the contractor will allocate a reliability target of 99.8% to each. The overall system will then have a reliability of 99.0%, calculated by multiplying the individual reliabilities of A, B, C, D and E. The difference of 0.2% between the target reliability of 98.9% and the predicted reliability of 99.0% will be the contractor's margin.
However, what if subcontractor B's equipment cannot achieve a reliability of 99.8% due to technological constraints? In this case, even if the contractor allows subcontractor B to provide equipment with a reliability of 99.6%, the total reliability will still be 98.81%, which satisfies the target service reliability because of the margin the contractor has.

Traceability and change control

Traceability matrix

< Traceability >

All requirements will necessarily be tracked in terms of the following aspects over the project lifecycle:

- How has each requirement been decomposed?
- How has each requirement been actualised?
- How has each requirement been changed?
- How has each requirement been verified and validated?

From the client's standpoint, the above questions should be answered to accept a system. A traceability worksheet or database can provide all the necessary information needed to manage requirements and monitor their status. Traceability provides the ability to describe how a requirement has been processed in both downward and upward directions.

Near the end of the project, a compliance matrix needs to be prepared, as reviewing the traceability worksheet or database can be complicated for a client checking the deliverables. The compliance matrix can be produced by capturing the original requirements and documenting the evidence found through V&V process. The compliance matrix is used by the client to verify whether deliverables have been developed based on the requirements and to ensure all requirements are met.

< Traceability matrix >

The requirements traceability matrix is a worksheet (or database) that demonstrates the relationship between requirements and other artifacts. Requirements managers use it to allocate requirements to each development team and to prove that requirements have been fulfilled. The requirements traceability matrix consists of the following fields:

- Original requirements: requirements ID, chapter number, description
- Decomposed system requirements: requirements ID, criteria
- Project lifecycle where a requirement is verified or validated: design, manufacture, installation, testing and commissioning (T&C)
- Development teams or disciplines (requirement owners)

Generally, Excel spreadsheets are used for developing the requirements traceability matrix; however, specialised requirements management software, such as DNG (Doors Next Generation), can also be employed. These tools provide methods to define, trace, analyse, and manage requirements systematically. Requirements managers can use both the specialised software and Excel spreadsheets interchangeably by exporting and importing requirements data between them because the two types of software are complementary:

- Excel spreadsheets: easy to manipulate, rapid in data control, and straightforward to share with stakeholders
- Requirements management system: suitable for large databases and offering various functions

< Requirements baseline and management process >

When the entire set of the project requirements is established, it should be approved by the client or, at the very least, reported to them to establish the requirements baseline officially. This is a crucial step for a system integrator to lead the systems engineering activities of all participants, so it should be included as one of the requirements processes.

Once the requirements baseline is officially distributed to the project participants, the requirements change control procedures should also be communicated to them. This ensures that all requirement changes can be tracked. Figure 25 illustrates how requirements are processed and controlled throughout the project lifecycle.

Figure 25 – Requirements management process

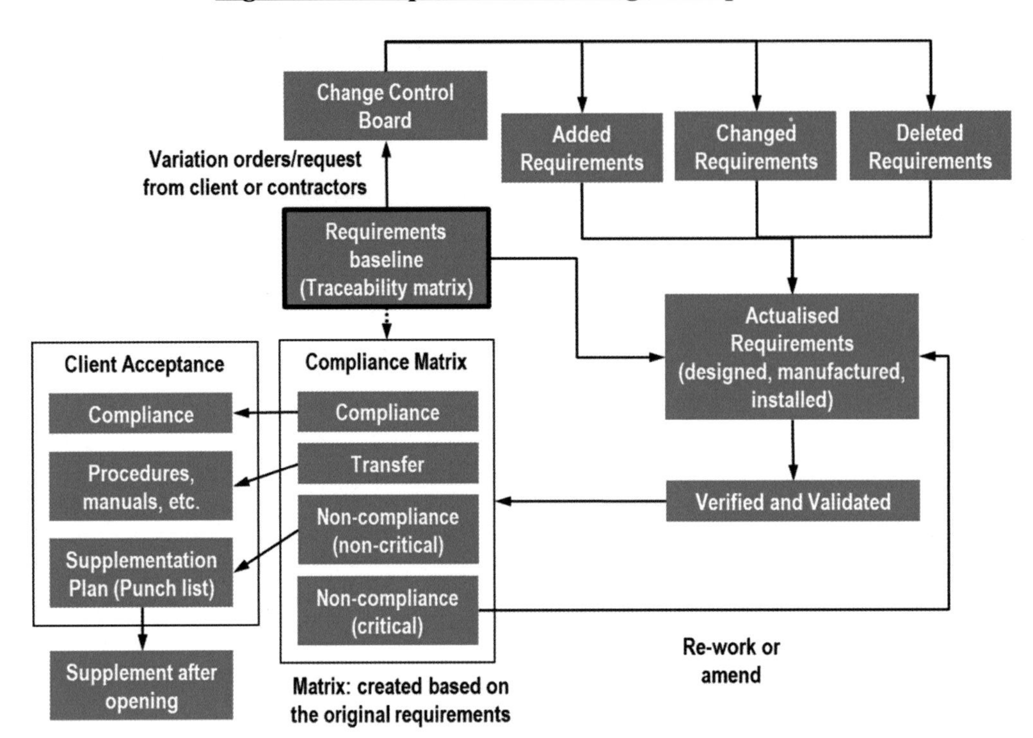

Requirements change control

As the project progresses, requirements may be deleted, added, merged, or split for various reasons, as illustrated in Figure 26. By the end of the project, the final requirements may differ somewhat from the original ones. Change control is essential to ensure that any modifications to requirements are systematically evaluated, approved, and documented. This process prevents scope creep, minimises project risks, and ensures that all stakeholders are aware of and aligned with the changes. Effective change control maintains project stability, keeps development on track, and ensures that changes are implemented in a controlled and predictable manner.

Figure 26 - Requirements changes

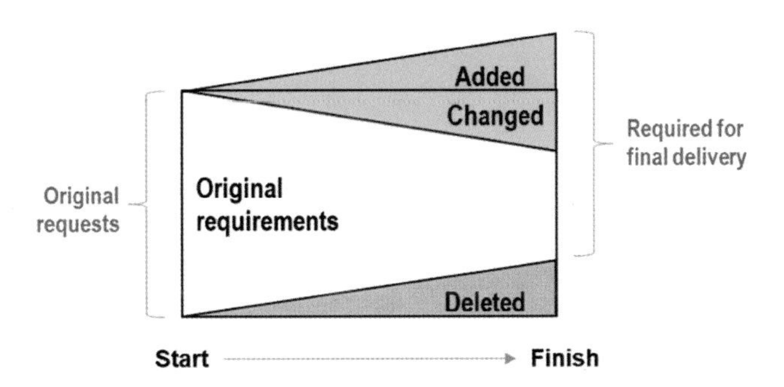

< Strategies and tips to control requirement changes effectively >

Here are some tips to help a requirements manager efficiently control changes of requirements:

- Process changes officially: The requirements change control process should be established first and then shared with stakeholders.
- Prepare Change Log Sheets: Create change log sheets that include as much change information as possible, as this information will be used for verification and validation. For example, when using Excel sheets, columns should represent fields where all requirement information are recorded. It is essential to derive as many information types or requirements attributes as possible and record them in the fields.
- Avoid changing requirements descriptions: It is better not to change the descriptions of the requirements; instead, delete the existing requirement and create a new changed requirement with a new ID number. Changing the description requires version control, which adds unnecessary work. When modifying a requirement, the requirements manager should delete the original requirement and creating a new requirement with a new ID number, leaving a comment such as "This requirement is replaced with ID xxx."
- Do not reuse deleted requirement ID numbers: Reusing ID numbers of deleted requirements can cause confusion among engineers.

Requirements change procedure

To control changes, the Change Control Board (CCB) is typically run by the client, as changes often involve schedule and/or cost adjustments, which are contractual matters.

The requirements change procedure is a structured process for handling modifications to project requirements. This systematic approach minimises disruption and ensures that all relevant parties are informed and aligned with the project scope and target. It typically involves the following steps:

- Submitting a change request: A formal request is made to propose changes to the existing requirements.
- Assessing impact: The impact of the change on scope, schedule, and resources is evaluated.
- Obtaining approval: Approval is sought from the client with the help of key stakeholders to ensure consensus on the proposed changes.
- Updating the project documents: Once approved, relevant documents, such as configuration sheets and logistics breakdown structure, are updated accordingly.

If terms or conditions of the contract are changed,

- The changed scope should be defined: Clearly articulate what the new scope entails.
- The impact on the schedule and cost should be analysed: Evaluate how the changes will affect the project timeline and budget.
- A claim expert should document the claim: Ensure that a claim expert formally documents any claims arising from the changes.

Figure 27 outlines the process by which a requirement can be changed.

Figure 27 - Detailed process of requirement changes

Phase	Change Control Board [Client & contractor]	Sub-contractors	Systems Engineering Team
Change request	Need to change	Need to change	Overall changes or changes relating to the interface between sub-contractors. Need to change Changes relating to a sub-contractor.
Analysis & Approval	Review changes in terms of schedule and costs Approve based on the reviews Issue the change	Analyse the impact on functions and performance	Analyse the impact on safety and RAM targets if needed
Execution		Perform design change	Apply changes to requirements, configuration, etc.

Verification and Validation (V&V)

Verification and validation (V&V) are the processes used for checking whether deliverables satisfy the requirements. These activities take place after requirements have been actualised into design and products.

Verification vs. validation

There is often confusion regarding the definitions of verification and validation. Generally, verification is defined as the process of ensuring that a product is developed correctly, while validation is the process of ensuring that the right product has been developed.
Some standards, such as EN 50126, suggest that validation is performed before handover, while verification is implemented at every project phase. However, validation may sometimes need to be conducted in the early phase, as it is generally easier to address non-compliances during the design phase than during installation. Consequently, other standards define V&V in various ways, including:

The PMBOK guide defines them in the 4th edition as follows:

- Verification: The evaluation of whether a product, service, or system complies with a regulation, requirement, specification, or imposed condition. It is often an internal process, in contrast with validation.
- Validation: The assurance that a product, service, or system meets the needs of the customer and other identified stakeholders. It often involves acceptance and suitability with external customers, contrasting with verification.

ISO 9001:2015 defines:

- Verification: Design verification is confirmation by examination and provision of objective evidence that the specified input requirements have been fulfilled. Verification activities include modelling, simulations, alternative calculations, comparison with other proven designs, experiments, tests, and specialist technical reviews.
- Validation: To ensure that the resulting product can meet the requirements for the specified application or intended use, where known. Design validation is like verification, except that the designed product should be checked under conditions of actual use.

In a company training textbook, verification and validation are introduced like:

- Verification: To check whether engineers have gone through the right pathway (as planned).
- Validation: To check whether engineers have arrived at the right destination to build a system.

The author prefers the following definitions:

- Verification: The evaluation of whether a deliverable meets regulations, requirements, and/or specifications.
- Validation: The assurance that a deliverable meets the operational needs of the user.

Verification & Validation (V&V) process

Here are some principles of V&V:

- Most international standards require that V&V should be done by experienced personnel.
- Criteria should be prepared before implementing V&V. (No criteria, no V&V activities.)
- A requirement can be judged to be satisfied only when relevant evidence is provided.
- Basically, when lower-level requirements are satisfied, relevant upper-level requirements can be considered satisfied because the lower-level requirements are components of the upper-level requirement. Figure 28 shows how upper requirements are satisfied. For instance, when requirements B2-1 and B2-2 are satisfied, requirement B2 can be judged to be satisfied. And when requirements Bl, B2, and B3 are satisfied, requirements B is judged to be satisfied.

Figure 28 - Integration

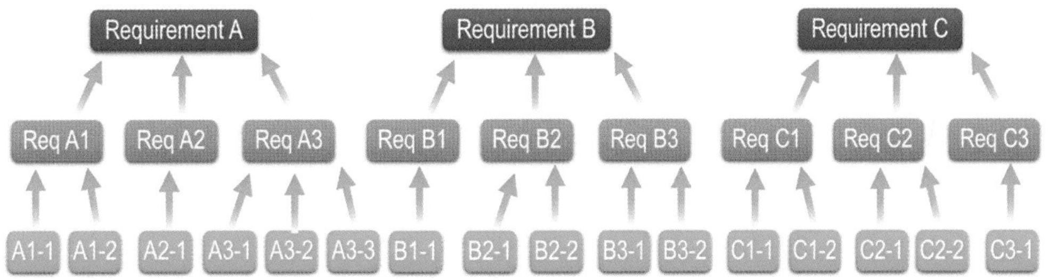

< V&V methodologies >

Verifier and validator can use the following methodologies in their activities:

- Inspection: Based on a visual or numerical review of the components, it relies on human senses or is verified by simple measurement and handling.
- Analysis: Prove theoretical conformity based on analytical evidence obtained using mathematical and probabilistic calculations, logical reasoning, modelling, and simulations under set conditions without intervention of submitted components.
- Demonstration: Used to show that submitted components work correctly based on operational and observed characteristics without physical measurements, usually with a selected set of activities to prove that the component's response to the stimulus is appropriate or that the operator can perform the assigned task using the component.
- Test: Quantitative validation of functional characteristics, measurable characteristics, operability, supportability, and performance with actual or simulated application of controlled conditions for submitted components.
- Simulation: Implemented on a model or physical model to demonstrate that certain elements and performances are consistent with the design.
- Sampling: Based on characteristic validation using samples; numbers, tolerances, and other characteristics must be specified, demonstrating that they are consistent with empirical feedback.
- Certificate review: A certificate is also evidence for verification and validation when its scope is fit for the criteria.

< Non-compliance >

At the end of the project (near the commercial opening), some non-compliance items may remain unresolved. The following classifications outline how non-compliance items are classified and dealt with:

- Critical non-compliance items: These items might affect the operation and require rework.
- Unimportant non-compliance items: These items do not affect the operation. If there is no time to fix them before the commercial opening, a schedule for supplementary work should be planned and implemented later.
- Transfer items: These items cannot be performed by the contractor, such as regular staff training. The contractor should prepare preliminary plans and hand them over to the client (or the operator).

RAM (Reliability, Availability, Maintainability)

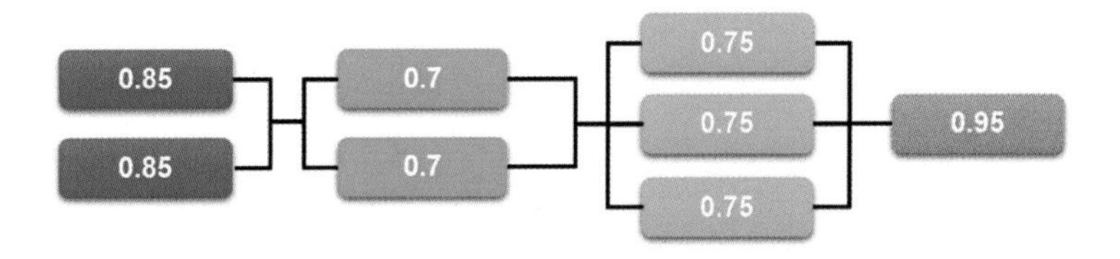

RAM definitions

RAM is an acronym for Reliability, Availability and Maintainability, commonly used in engineering to manage these aspects of a product or system. In general Systems Engineering, RAMS is also commonly used, which includes safety. RAM serves as a decision-making technique to optimise the availability of a system and reduce the failure costs.
To understand RAMS intuitively, the following questions can be used, assuming you have a car:

- Reliability – How often does your car fail? (These failures exclude those caused by external factors such as disasters or attacks, focusing instead on issues within the system and the interfaces between components.)
- Maintainability – How quickly can your car be repaired?
- Availability – Based on your car's reliability and maintainability, what is the probability that you can use it when you need it?
- Safety – Will anybody be injured or killed when your car fails? (Although RAM and safety are discussed concurrently, RAM is explained in this chapter, and safety will be addressed in the chapter on "SAFETY MANAGEMENT.")

From the standpoints of both operators and passengers, availability and reliability are among the most important quality indicators of rail service. If a railway line experiences frequent disruptions that cause delays, most passengers are likely to be dissatisfied, and operators may struggle to manage the resulting claims from passengers. This issue extends beyond mere claims; it impacts overall service reputation and customer trust. Ideally, a system should be perfect in many respects. However, developing a flawless system can be prohibitively expensive. Therefore, how much should we enhance the levels of availability and reliability? To answer, it is essential to understand RAM management techniques.

MTBF and MTBSAF

< MTBF >
MTBF is an abbreviation for "Mean Time Between Failures" and represents the average time between system failures. It is a key indicator for calculating reliability and availability of a system. In the railway industry, MTBF is typically expressed as a period, such as 20,000 hours or 10,000 hours. It is calculated

as follows (note that all units of time used for handling and calculating RAM data should be in "hours" to prevent calculation errors):

$$\text{MTBF (hr)} = \frac{\text{Total operation time}}{\text{Total number of failures}}$$

Imagine that a signalling system ran 18 hours per day and broke down 3 times in a week. Each outage took 2 hours to recover. MTBF is calculated as follows:

- Original total operation time: 126 hours = 18 hours x 7 days
- Total downtime: 6 hours = 3 times x 2 hours
- Total operation time: 120 hours = 126 hours – 6 hours
- MTBF: 40 hours = 120 hours / 3 times

When dealing with actual RAM data, the MTBF value can sometimes exceed several hundred thousand hours. In this case, there is no significant problem even if the downtime is ignored and not calculated.

< MTBSAF >

MTBSAF means "Mean Time Between Service Affecting Failures" and is used to calculate the impact of service-related failures. Not all subsystems or components that make up the entire rail system necessarily affect the rail services. Some equipment in the depot is a good example of non-service-affecting equipment. A wash plant is used for cleaning the body of a rolling stock, and even if it is out of order, a rail operator can continue the service. To estimate MTBSAF, components affecting the service should be identified. Generally, the following components are mainly considered as the items affecting the service:

- Rolling stock: engines, motors, doors, brakes, and TCMS (Train Control Management System)
- Signalling: interlocking, etc.
- Power supply: components in substations including transformers, and circuit breakers, etc.
- Telecommunication: backbone for transmission
- Station: screen doors
- Rail: S&Cs (Switches and Crossings)

MKBF and MCBF

There can be equipment whose availability should be calculated based on factors other than time, such as driving distance, cycle, frequency of use. Those numbers, rather than time, should be converted into time-based numeric data to be used for calculating availability, which is always a time-based ratio. MKBF (Mean Kilometres Between Failure) and MCBF (Mean Cycles Between Failures) are examples of indicators using non-timescale units.

< MKBF or MDBF >

MKBF means "Mean Kilometres Between Failures," and MDBF stands for "Mean Distance Between Failures." Generally, rolling stock has this type of data because MKBF is more reasonable – the further it travels, the higher the probability of failure. Since MKBF cannot be used for calculating availability, it should be converted into a time-based indicator.

Assume that a rolling stock has the operation conditions as follows:

- The MKBF for a rolling stock is 800,000 kilometres.
- The number of operations of each rolling stock is 4 round trips per day.
- The length of the line is 100 kilometres.
- The operation hours per day is 16 hours.

The calculation to convert MKBF into MTBF for a rolling stock is as follows:

- Operation kilometres per day: 800 kilometres = 100 kilometres x 2 x 4 round trips
- Kilometres per hour: 50 kilometres = 800 kilometres / 16 hours
- MTBF: 16,000 hours = 800,000 kilometres / 50 kilometres per hour
- Failure rate per hour: 0.00625% = 1 hour / 16,000 hours

< MCBF >

MCBF is an abbreviation for "Mean Cycles Between Failures" and is generally used for equipment such as train doors and AFCs (Automated Fare Collectors).

Imagine that the operation conditions and environment with AFCs are as follows:

- ▪There are four AFCs in a station. (two for entrance and two for exit)
- The MCBF for each AFC is 10,000 cycles.
- The number of passengers using the station is 3,200 per day. (2,800 boarding and 1,400 alighting)
- The total operation hours of the station are 14 hours.

The calculation to convert MCBF into MTBF for each AFC used for entrance is as follows:

- Number of cycles of each AFC used for entrance (assume that two AFCs share the same number of passengers): 1,400 cycles = 2,800 passengers (boarding) / 2

- Cycles per hour per AFC: 100 cycles = 1,400 cycles / 14 hours
- MTBF: 100 hours = 10,000 cycles / 100 cycles per hour
- Failure rate per hour: 1% = 1 hour / 100 hours

If one of the two AFCs fails, the other can still be used. In this case, the total reliability should be calculated using a parallel configuration. The method for calculating total reliability in either parallel or series configuration is addressed in the chapter titled "Configuration".

On the other hand, the MTBF for AFCs used for exit can be calculated as follows:

- Number of cycles of each AFC used for exit (assume that two AFCs share the same number of passengers): 700 cycles = 1,400 passengers (alighting) / 2
- Cycles per hour per AFC: 50 cycles = 700 cycles / 14 hours
- MTBF: 200 hours = 10,000 cycles / 50 cycles per hour
- Failure rate per hour: 0.5% = 1 hour / 200 hours

It might seem strange that the MTBFs of AFCs for entrance and exit differ, but the discrepancy arises because the AFCs for entrance are used more frequently due to the higher number of passengers boarding. This highlights the importance of verifying values when converting MCBF to MTBF. Higher usage results in a lower MTBF, which is why checking the conversion thoroughly is essential to ensure the accuracy of the calculations.

MTTR and MART

< MTTR >

MTTR, which stands for "Mean Time To Repair" (or sometimes "Mean Time To Restore" from a rail operator's perspective) is calculated using the following equation:

$$\text{MTTR (hr)} = \frac{\text{Total downtime}}{\text{Total number of failures}}$$

Let's suppose that a power supply system ran 20 hours per day and broke down 4 times in a week. The first two outages took 1.5 hours, the third 1 hour, and the fourth 2 hours to recover. MTTR is calculated as follows:

- Total repairing time: 6 hours = 1.5 hours x 2 times + 1 hour + 2 hours
- MTTR: 1.5 hours = 6 hours / 4 times

As defined by IEC 61508 and ISO TR 12489, a maintainer will experience the process of restoration, each step will take time. The times relevant to the steps are called as follows:

Figure 29 - Restoration process

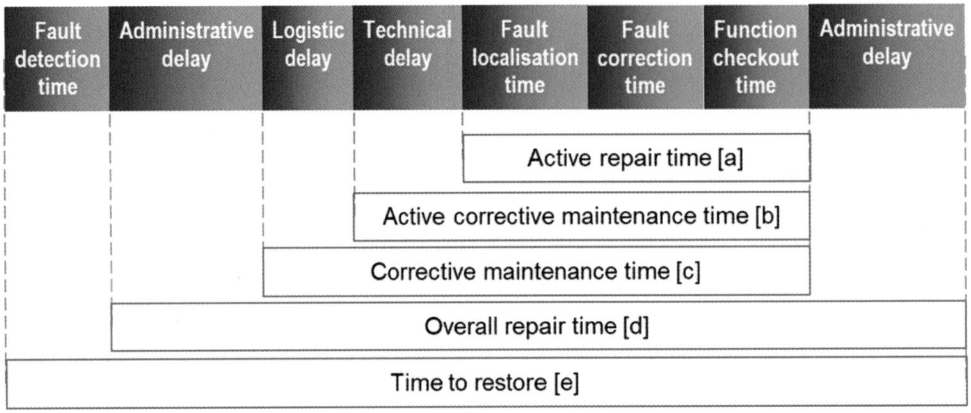

Fault detection time
Administrative delay
Logistic delay
Technical delay
Fault localisation time
Fault correction time
Function checkout time
Administrative delay
Active repair time [a]
Active corrective maintenance time [b]
Corrective maintenance time [c]
Overall repair time [d]
Time to restore [e]

According to the definition described in IEC 61508, MTTR is actually "Time to restore [e]" covering all maintenance steps. However, some aspects are not directly related to real maintenance activities. For example, logistic delay depends on the location and number of maintenance offices, as systems and components in a railway are dispersed over long distances. Therefore, such time and delay should not be included when calculating actual maintainability.

< MART >
MART is the abbreviation of "Mean Active Repair Time" defined by IEC 61508 and is the sum (active repair time [a]) of fault localisation time, fault correction time, and function checkout time, as Figure 29 shows. MART is more reasonable metric than MTTR because other times and delays included in MTTR are not relevant to the actual repair time but to administrative or logistic activities. It is preferred in estimating actual maintenance time, and, therefore, suppliers provide MART values for maintainability calculation. On the other hand, MTTR is more commonly mentioned depending on industry practices when performing RAM tasks. Thus, MTTR will be referred to as MART in this book.

Configuration

< Series configuration >

Series configuration of components means that if any component in a system fails, the whole system is regarded as failed. This is a typical composition of a system. If each component in a system performs a different function and the failure of single component affects the performance of the whole system, their connections are regarded as serial connections. It is like series connection of batteries; if battery A or B fails, the bulb does not light up.

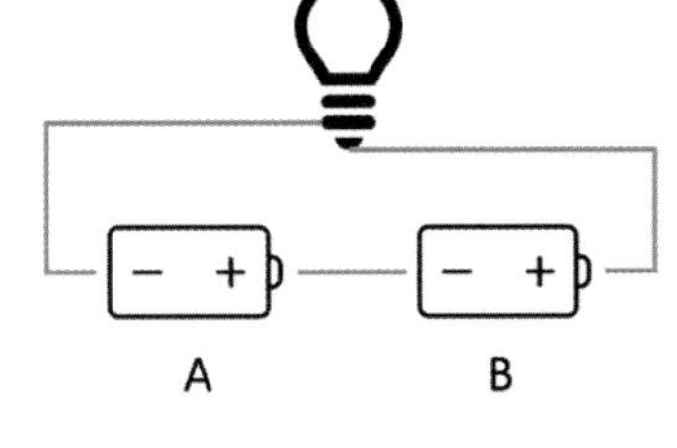

< Parallel configuration >

Parallel configuration is a configuration in which a system has one or more redundant components that would perform a required function. Even if any of them fails, whether they are actively or passively operating, the system still can operate. It is like parallel connection of batteries; if battery A or B fails, the bulb still can light up.

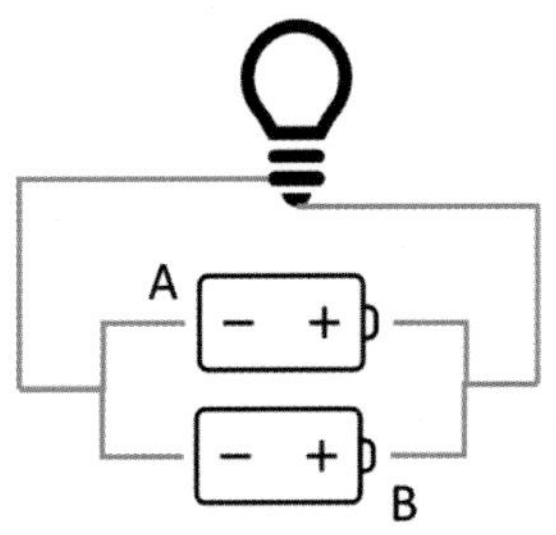

< "k out of n" configuration >

If grouped components in a system perform the same function and a number (k) of components out of the total number (n) of them are required for success, their configuration is regarded as "k out of n" configuration.

RAM calculations

Failure rate calculation

A failure can be defined as a status where a required function cannot be performed due to issues within the system and/or problems at the system interface. Failure rate is the frequency of component failure per unit time (hour). Here is the equation:

$$\text{Failure rate (hr)} = \frac{1}{\text{MTBF}}$$

Failure rate is one of the most important indicators for handling RAM data, and it is also used to calculate the quantity of required spare equipment or devices. It is usually denoted by the Greek letter λ (Lambda) in many papers. Since the failure rate is the inverse of MTBF, it can be derived from the following equation:

$$\text{Failure rate (per hour)} = \frac{\text{Number of failures}}{\mathit{Total\ operation\ time\ (hours)}}$$

< Overall failure rate >

When considering redundancy design with multiple units, the overall failure rate equation approximations, according to the "Reliability Toolkit: Commercial Practices Edition" (Reliability Analysis Centre), are as follows:

< With Repair >

When a failed unit is repaired immediately upon failure, the overall failure rate for multiple units in a parallel configuration for redundancy design can be approximated using the following equations:

		With Repair (This should not be applied if delayed maintenance is used)
Two active on-line units with different failure and repair rates. One of two required for success.	Total $\lambda_{1/2}$	$= \frac{\lambda_a \times \lambda_b \times (\lambda_a + \lambda_b + \mu_a + \mu_b)}{(\lambda_a + \lambda_b) \times (\mu_a + \mu_b) + \mu_a \times \mu_b}$
All units are active on-line with equal unit failure rate. (n-q) out of n required for success.	Total $\lambda_{(n-q)/n}$	$= \frac{n! \times \lambda^{q+1}}{(n-q-1)! \times \mu^q}$
One standby off-line unit with n active on-line units required for success. Off-line spare assumed to have a failure rate of zero. On-line units have equal failure rates.	Total $\lambda_{(n/n)+1}$	$= \frac{n \times [n \times \lambda + (1-P) \times \mu] \times \lambda}{\mu + n \times (P+1) \times \lambda}$

where:

- λ: failure rate of an individual on-line unit (failure/hour)
- $\lambda(x/y)$: the effective failure rate of the redundant configuration where x of y units required for success
- μ: repair rate (μ=1/Mct, where Mct is the mean corrective maintenance time in hours)
- n: number of active on-line units
- q: number of on-line active units which are allowed to fail without system failure
- n-q: k (when applying "k out of n")
- P: probability switching mechanism will operate properly when needed (P=1 with perfect switching)

< Without Repair >

When a failed unit is repaired after the operation closing, the overall failure rate for multiple units in a parallel configuration for redundancy design can be approximated using the following equations:

		Without Repair
Two active on-line units with different failure and repair rates. One of two required for success.	$\text{Total } \lambda_{1/2}$	$= \frac{\lambda_a \times \lambda_b \times (\lambda_a + \lambda_b)}{\lambda_a^2 + \lambda_b^2 + \lambda_a \times \lambda_b}$
All units are active on-line with equal unit failure rate. (n-q) out of n required for success.	$\text{Total } \lambda_{(n-q)/n}$	$= \frac{\lambda}{\sum_{i=(n-q)}^{n} \frac{1}{i}}$
One standby off-line unit with n active on-line units required for success. Off-line spare assumed to have a failure rate of zero. On-line units have equal failure rates.	$\text{Total } \lambda_{(n/n)+1}$	$= \frac{n \times \lambda}{P + 1}$

Maintainability calculation

The overall MTTR of elements in a series configuration is calculated by taking the weighted average of the MTTRs of the individual elements, with the weighting based on the failure rate of each element. This means that elements with higher failure rates contributes more significantly to the overall MTTR. The equation for calculating the overall MTTR of elements in series configuration is:

$$Overall\ \text{MTTR}_{\text{series}} = \frac{\text{F.r}_1 \times \text{MTTR}_1 + \text{F.r}_2 \times \text{MTTR}_2 + \cdots + \text{F.r}_n \times \text{MTTR}_n}{\text{F.r}_1 + \text{F.r}_2 + \cdots + \text{F.r}_n}$$

The overall MTTR of elements in a parallel configuration for redundancy design is calculated as follows:

- In case that each unit has different MTTR:

$$\frac{1}{\text{Overall MTTR}_{\text{parallel}}} = \frac{1}{\text{MTTR}_1} + \frac{1}{\text{MTTR}_2} + \cdots + \frac{1}{\text{MTTR}_n}$$

- In case that each unit has equal MTTR:

$$\textit{Overall}\ \mathrm{MTTR}_{(\mathrm{k\ out\ of\ n})} = \frac{\mathrm{MTTR}}{\mathrm{n-k+1}}$$

Reliability calculation

Reliability is the probability that a system or component will perform its function under stated conditions for a specified period. It can be calculated using the failure rate as follows (where "e" is the base of the natural logarithm, approximately 2.71828):

$$\text{Reliability (t)} = e^{-\text{failure rate} \times \text{time } (\textit{hours})}$$

When determining the volume of time in the above equation, it is recommended to use the duration of the RAM demonstration. If the duration of the RAM demonstration is one year, it will be 8,760 hours (= 24 hours X 365 days).

Reliability is also an important indicator used to calculate the overall reliability of various types of component configurations, including not only series and parallel configurations but also "k out of n" type.

< Series configuration >
Imagine that there are computer components such as CPU, LAN adapter, or USB interface. If any of them fails, the computer is regarded as failed. Even though they are not directly connected as a series, we consider them as a series configuration to calculate the overall reliability. The overall reliability formula for series configuration is:

$$\textit{Overall}\ \mathrm{Reliability}_{\mathrm{series}} = \mathrm{R}_1 \times \mathrm{R}_2 \times \mathrm{R}_3 \times \cdots \times \mathrm{R}_\mathrm{n}$$

If there is a system consisting of three components with reliabilities of 0.98, 0.99, and 0.97 respectively in series configuration, the overall reliability can be calculated as follows:

$$0.941094 = 0.98 \times 0.99 \times 0.97$$

< Parallel configuration >

If grouped components in a system perform the same function and the failure of a single component does not affect the performance of the whole system, their connections are regarded as parallel connections. The overall reliability formula for parallel configuration is:

$$\mathit{Overall}\ \text{Reliability}_{\text{parallel}} = 1 - (1 - R_1) \times (1 - R_2) \times (1 - R_3) \times \cdots \times (1 - R_n)$$

When there is a system consisting of three components with reliabilities of 0.93, 0.92, and 0.94 in parallel configuration, the overall reliability can be calculated as follows:

$$0.999664 = 1 - (1 - 0.93) \text{ x } (1 - 0.92) \text{ x } (1 - 0.94)$$

< "k out of n" configuration >

If there is a system consisting of five devices, and three out of the five are required for success, this configuration is referred to as a "3 out of 5" configuration. The formula for calculating the overall reliability in a "k out of n" configuration, where all units have equal MTBF and MTTR, is as follows:

$$\text{Overall Reliability}_{\text{k out of n}} = \frac{n!}{(n-k)! * k!} * R^k * (1-R)^{n-k} + \frac{n!}{(n-(k+1))! * (k+1)!} * R^{k+1} * (1-R)^{n-(k+1)} + \cdots + \frac{n!}{(n-n)! * n!} * R^n * (1-R)^{n-n}$$

where:

- n is the total number of elements.
- k is the minimum number of functioning elements required for success.
- R is the reliability of each individual component.

Even though there is a simpler expression for calculating the overall reliability in "k out of n" configuration as shown below, the author prefers to use the formula above to help readers understand the concept more intuitively. Additionally, the formula above is more helpful to Microsoft Excel users.

$$R_s(k, n, R) = \sum_{r=k}^{n} \binom{n}{r} R^r (1-R)^{n-r}$$

If there is a system with seven components, each having a reliability of 0.95, and at least five of them must operate for success, the calculation is for the overall reliability using the "k out of n" configuration formula would be as follows:

$$\text{Overall Reliability}_{k\ \text{out of}\ n} = \frac{7!}{(7-5)! \times 5!} \times 0.95^{5} \times (1-0.95)^{7-5} + \frac{7!}{(7-(5+1))! \times (5+1)!} \times 0.95^{(5+1)} \times (1-0.95)^{7-(5+1)} + \frac{7!}{(7-(5+2))! \times (5+2)!} \times 0.95^{(5+2)} \times (1-0.95)^{7-(5+2)}$$

$$\text{Overall Reliability}_{k\ \text{out of}\ n} = 0.0462350 + 0.25728216 + 0.69833730 = 0.99624296$$

When calculating the reliability of a system consisting of many parts which have the same reliability (MTBF), both the formulas for parallel configuration and "k out of n" configuration yield the same overall reliability. The formula for parallel configuration is applied only when each part has different reliability, and only one needs to or will operate out of them.

< Overall reliability of all train sets >

Usually, railway service providers (operators) have train sets for operation and spare ones for in case of failure. In this case, it is not simple to determine the overall reliability of the total number of train sets.

How should the total reliability be calculated if there are ten train sets with reliabilities of 95% for operation and two spare train sets on a particular line? When one train set is on standby to run in case one of the trains in operation requires regular maintenance, and the other is on standby to run in case a train currently running fails, the following steps seem reasonable to calculate the overall reliability:

- Step 1 – Estimate the mean time to maintenance.
- Step 2 – Calculate "k out of n" on available train sets.
- Step 3 – Apply linear interpolation method.

To calculate the overall reliability, the first step is to check the maintenance schedule and period in depot. Suppose that the average maintenance time for the line is a total of 9 months per year as follows:

- On average, one of the train sets is under heavy maintenance for 6 months per year.

- One of the train sets is under scheduled maintenance for 2 months per year.
- One of the train sets is under unscheduled maintenance (due to failures) for 1 month per year.

This means that one of the two spare train sets should be in operation for 9 months per year; thus, reliability for "10 out of 11" can be applied as follows:

$$\text{Overall Reliability}_{k\ out\ n} = \frac{11!}{(11-10)! \times 10!} \times 0.95^{10} \times (1-0.95)^{11-10} + \frac{11!}{(11-(10+1))! \times (10+1)!} \times 0.95^{(10+1)} \times (1-0.95)^{11-(10+1)}$$

$$\text{Overall Reliability}_{\text{k out of n}} = 0.32930532 + 0.5680009 = 0.89810541$$

On the other hand, one of the two reserved train sets is on standby for 3 months per year, and reliability for "10 out of 12" for those 3 months should be calculated as well as follows:

$$\text{Overall Reliability}_{k\ out\ of\ n} = \frac{12!}{(12-10)! \times 10!} \times 0.95^{10} \times (1-0.95)^{12-10} + \frac{12!}{(12-(10+1))! \times (10+1)!} \times 0.95^{(10+1)} \times (1-0.95)^{12-(10+1)} + \frac{12!}{(12-(10+2))! \times (10+2)!} \times 0.95^{(10+2)} \times (1-0.95)^{12-(10+2)}$$

$$R_{all} = 0.09879159 + 0.34128006 + 0.54036009 = 0.98043174$$

The calculation is as follows:

- Difference between them: 0.08232633 = 0.98043174 – 0.89810541
- Contribution of 3 months (25% of a year): 0.02058158 = 0.08232633 x 25%
- Overall reliability: 91.86% = 89.81% + 2.05%

Of course, the overall reliability calculated using the linear interpolation method does not indicate the exact number, but it is more reasonable than ignoring the 3 months. Readers are encouraged to suggest more reasonable methods.

Availability calculation and RAM target

Availability is the probability of performing a required function under the given environment for a specified time. This time usually refers to operation time. In fact, availability is the ratio of uptime to total operating hours (uptime plus downtime). The calculation of availability is as follows:

$$\text{Availability} = \frac{\text{Total uptime}}{\text{Total uptime} + \text{Total downtime}}$$

To help understand the relationship between uptime, downtime, and operation time from another perspective, we can consider the following formula:

$$\text{Uptime} + \text{Downtime} = \text{Total operation time}$$

Considering the above equation, availability can be expressed as:

$$\text{Availability} = \frac{\text{Total operation time} - \text{Total downtime}}{\text{Total operation time}}$$

However, the above formula can only be used by rail operators. During the development phase, the following formula is generally used because suppliers only provide the MTBF and MTTR data:

$$\text{Availability (\%)} = \frac{\text{MTBF}}{\text{MTBF}+\text{MTTR}} \times 100$$

An example for availability calculation is explained in the chapter titled “RAM demonstration”.

< RAM target >

RAM targets are the minimum levels to be achieved in terms of reliability, maintainability, and availability. A RAM target can be found as a key requirement, such as KPI, in the client’s requirements, which are generally described in contractual documents. The targets are usually expressed as numbers and derived from operational performance target as follows:

- Reliability – MTBF (e.g., 90,000 hours)
- Maintainability – MTTR or man-hour (e.g., 4 man-hour)
- Availability – percentage (e.g., 99.8%)

If no RAM target is specified in the contractual documents, but the contractor is required to establish one, the contractor should establish the target by analysing the RAM performance of comparable projects.

< Overall availability vs. operational availability >

Two types of formula are broadly used in the railway industry to establish operational (or service) availability targets, as described in chapter titled “Service availability”: the schedule-based formula and the reliability-based formula. Of these, the calculation method for the reliability-based is nearly identical to that for overall availability discussed in this chapter. The only difference lies in the selection of components. For overall availability, the MTBF and MTTR of all components are used, whereas for operational (or service) availability, only components associated with rail service-affecting delays are selected. The formula is:

$$\text{Overall availability} = \frac{MTBF_{overall}}{MTBF_{overall} + MTTR_{overall}}$$

This formula is used to monitor and control only fault events in a system that cause service-affecting failures, rather than the system’s intrinsic reliability.

As mentioned in the earlier chapter, the maintenance offices are dispersed along the long railway line, resulting in significant logistical time required to access the point of failure. Therefore, this logistical time should not be included in the total repair time. The distance between the maintenance offices is not considered a system issue but rather a matter of maintenance policy. Consequently, MART (Mean Active Repair Time) is preferred in railway services as follows:

$$\text{Service Availability} = \frac{MTBSAF}{MTBSAF + MART}$$

RAM Management Plan

For a RAM manager at the top level in the organisational structure of railway projects, it is essential to develop a RAM Management Plan that will guide participants in communicating, cooperating, and developing documents associated with RAM tasks. The plan should include, at a minimum, the following:

- Scope of work
- System description
- RAM targets or requirements, including RAM acceptance criteria
- Organisation chart
- Roles and responsibilities
- Methodologies and RAM process, including verification and validation processes
- Programme of RAM assurance activities to be undertaken throughout the project lifecycle

Here are some references for RAM planning:

- EN 50126: Railway applications – Specification and demonstration of Reliability, Availability, Maintainability and Safety (RAMS)
- EN 50128 – Railway applications (Signalling and Software)
- EN 50129: Railway applications – Communication, signalling and processing systems – Safety related electronic systems for signalling
- Reliability Toolkit: Commercial Practices Edition
- Electronic Reliability Design Handbook – Military Handbook (Department of Defence, US)

Scope of work

The scope of RAM management work should be clearly defined in the RAM management plan. As mentioned in the previous chapter, various teams at different levels work according to their assigned roles and responsibilities. A clear and detailed scope of work will help avoid "unnecessary tasks".

To develop the "Scope of work," the following input will be required:

- Contractual documents
- Project-wide RAM plan (if applicable)

- System definition
- Project plan

The activities to develop the "Scope of work" using the above input will include:

- Reviewing the requirements in the contractual documents and translating them into RAM objectives and targets
- Reviewing the higher-level RAM management plan

< System description >

The system description provides a detailed explanation of the system to be developed in terms of the work scope. The following inputs will be required:

- Contractual documents
- Project-wide RAM plan (if applicable)
- System definition

The activities to develop the "System description" using the above inputs will include:

- Drawing the system to be developed
- Identifying the system boundary
- Identifying the system interfaces

The system boundary should be defined in terms of:

- Physical boundary – e.g., location, quantity
- Interface boundary – interface points
- Functional boundary – including interface functions
- Hardware & software configuration and system capability

RAM requirements and apportionment

All RAM activities will be conducted based on RAM-related requirements derived from contractual documents, as well as relevant laws, regulations, standards, specifications, and other pertinent sources. These requirements should be outlined in the RAM Management Plan.

One of the key requirements among the RAM requirements is RAM targets, which should be allocated to each discipline (or subsystem, element, etc.) – a process referred to as RAM apportionment. The RAM apportionment allocated to each discipline represents its level of contribution to the overall RAM target.

< Roles and responsibilities >
The roles and responsibilities of the following participants should be defined as follows:

- Project Director
- Project Manager
- Systems Engineering Manager
- RAM Manager
- Safety Manager
- Requirements Manager
- Verification and Validation Manager
- RAMS Managers of subsystem suppliers

To define the "Roles and responsibilities", the following inputs will be included:

- Contractual documents
- Project-wide RAM plan (if applicable)
- Project plan

The activities to develop the "Roles and responsibilities" using the above input will include:

- Identifying responsibilities for all key personnel, such as project manager, RAM manager, and others.
- Outlining the qualifications required for each key personnel.
- Identifying independences among roles.
- Integrating the RAM management team into the overall organisation.
- Briefly describing the interaction between parties involved.

The outputs from the above activities will include:

- Organisation chart
- RAM organisations with roles & responsibilities
- Roles and responsibilities for each personnel
- Competencies required for each personnel

- Independence (e.g., reviewer, approver, etc.)

RAM deliverables programme

RAM deliverables programme is a documented schedule for RAM deliverables. The RAM programme can take the form of a Gantt chart or timetable. It enhances communication among personnel involved in RAM-related deliverables.
However, it may be necessary to update the schedule frequently as the project progresses and the project schedule changes. The author has experienced spending considerable time revising the programme of the RAM plan due to these changes. In the author's opinion, it is not advisable to include the RAM deliverables programme within the RAM Management Plan. Instead, it is recommended to maintain a separate document that contains the overall systems engineering schedule. If the client requires the RAM plan to include the schedule, it is suggested to develop the schedule using phases – such as design phase, installation phase, and testing and commissioning (T&C) phase – rather than specific dates, which will reduce the frequency of updates.

The inputs for developing the "RAM deliverables programme" are as follows:

- Contractual documents
- Project-wide RAM plan (if applicable)
- Project plan & milestone programme

The activities will include:

- Reviewing project-wide milestones
- Identifying the activities related to RAM management
- Determining the timeframe for exchanging information with subcontractors and suppliers
- Planning verification and validation activities before the end of each phase.

Presumptions/prerequisites & constraints

When developing the RAM Management Plan, many aspects may not yet be clearly defined due to the project's lack of maturity. Therefore, presumptions (or assumptions) & prerequisites should be documented in the management plan as unresolved items. Additionally, constraints, such as environmental constraints, should also be mentioned.

Examples of assumptions may include:

- Potential design options
- Operation and maintenance (O&M) assumptions
- Potential suppliers

Examples of constraints may include:

- Properties of specific equipment and products
- Environmental constraints
- Provided equipment

The inputs will include:

- Contractual documents
- Project-wide RAM plan (if applicable)
- Project plan
- System definition
- Safety-Related Application Conditions (SRACs) of products

The activities will involve:

- Developing the general assumptions to be applied in the project
- Obtaining agreement from the project team on the prepared assumptions

RAM management process and subcontractors management

< RAM management process >

RAM management activities should be described in the plan to assist other engineers participating in the project in understanding how to cooperate on RAM management tasks. Detailed activities can be outlined as follows (including a graphical workflow is highly preferable):

- Inputs
- Activities
- Outputs

< Subcontractors management >

If there are subcontractors or suppliers to be managed for RAM activities, the plan should include the following:

- Their roles and responsibilities regarding RAM activities
- Competencies required for RAM engineers
- Configuration management and change control process

The inputs for developing the "Subcontractors management" chapter are as follows:

- Contractual documents
- Project-wide RAM plan (if applicable)
- Project plan
- System definition and work breakdown
- Initial apportionment of RAM requirements
- Subcontractor agreements (if applicable)

< Verification and validation >

It is essential to prepare RAM acceptance criteria for RAM verification and validation (V&V). Typically, RAM acceptance criteria are quantitative metrics, such as MTBF and MTTR. For V&V activities, it is recommended to use FRACAS or DRACAS to record all failures.

The inputs are as follows:

- RAM requirements or targets
- Project-wide RAM plan (if applicable)
- Project plan
- Standards and guidance
- Sources of RAM data
- RAM demonstration test tools and test environment

The activities are as follows:

- Translating requirements into demonstrable targets
- Agreeing with client on the demonstration testing process
- Reviewing relevant standards
- Developing pass/fail criteria

RAM techniques

The ultimate purpose of RAM management is to enhance cost-effectiveness through design optimisation, as illustrated in Figure 30. By employing RAM techniques, system design can be optimised, leading to numerous benefits, ultimately resulting in improved cost-effectiveness. Optimising design in terms of RAM level involves trading off between construction costs and reliability levels, as depicted in Figure 31.

Figure 30 - Design optimisation & cost effectiveness

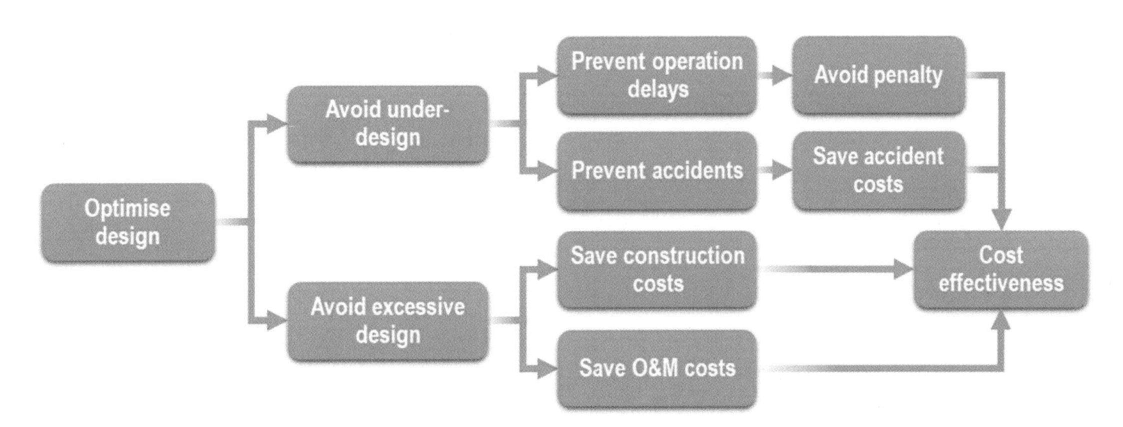

Figure 31 - Optimal RAM level

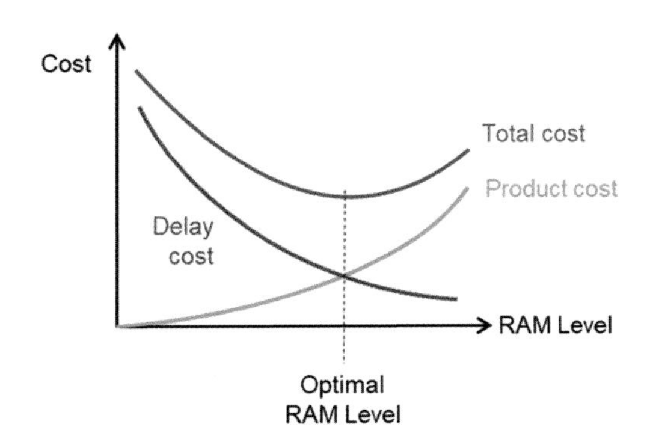

To achieve cost-effectiveness, the following RAM activities will be conducted:

- Gather the RAM-related requirements
- Establish RAM target
- Study the characteristics of systems and subsystems to identify components a high disruption potential to service
- Allocate RAM target apportionment to each subsystem based on component analysis
- Perform FMECA (Failure Modes Effects and Criticality Analysis) and other RAM analysis using RAM data provided by subsystem providers
- Adopt RAM elements through RAM analysis (RAM prediction)
- Verify and validate the evidence
- Develop arguments to demonstrate compliance with RAM requirements
- Obtain acceptance from the project owner
- Demonstrate whether the system meets the RAM target during the operation phase

Among the RAM activities mentioned above, RAM analysis activities are particularly crucial. RAM analysis involves quantitatively calculating the reliability, maintainability, and availability of a single subsystem or the entire system through techniques such as FMECA, FTA (Fault Tree Analysis), and RAM prediction. This process aids in identifying potential issues early in the design phase, allowing for necessary adjustments to be made before the system is fully implemented.

RAM allocation

The overall RAM target is typically defined in the contractual documents as a part of KPIs (Key Performance Indicators). RAM target allocation refers to the initial breakdown of the overall RAM target, with apportionments assigned to subsystem suppliers. For effective RAM management, the integrator should have a RAM margin. For example, if the RAM target is the availability of 99.2%, the overall availability of the individual apportionments should be at least 99.5%. The difference between them constitutes the integrator's RAM margin, which can be utilised if a subsystem supplier fails to meet the assigned RAM target.

Before allocation, it is essential to review the service availability formula outlined in the "Service availability" chapter of the "OPERATION PERFORMANCE" section, in accordance with contractual requirements. If the contract stipulates that service availability must be computed based on the trip schedule, the allowed number of delayed trips will be allocated. Conversely, if service availability is calculated using MTBF and MTTR, the allowed number of service failures will be allocated. In some projects, reliability and maintainability are allocated separately instead of allocating availability.

In the O&M (Operating and Maintaining) phase, FRACAS (Failure Reporting, Analysis, and Corrective Action System) will be utilised to assess whether the RAM target have been achieved. This technique also identifies the contributors to service delays and failures.

Figure 32 illustrates examples of availability allocation in various railway projects, where each subcontractor (e.g., rolling stock, signalling, etc.) is assigned a specific availability target. A client may require MTBSAF and MTTR as separate metrics. When the systems engineering team allocates the overall availability, it is essential to secure an availability margin to account for instances where a subcontractor fails to deliver components that meet the assigned RAM target. In the case of project C, the margin is set to zero, which is not a desirable practice for RAM allocation.

Figure 32 - Examples of availability allocation

	Project A	Project B	Project C	Project D	Project E
Rolling stock	**99.96%**	**99.99%**	**99.54%**	**99.87%**	**99.50%**
Signalling	**99.94%**	**99.97%**	**99.80%**	**99.90%**	**99.94%**
Power supply	**99.94%**	**99.97%**	**99.92%**	**99.97%**	**99.94%**
Communication	**99.93%**	**99.96%**	**99.92%**	**99.98%**	**99.93%**
PSD	**99.96%**	**99.97%**	**99.92%**	**99.95%**	**99.96%**
Availability [a]	99.73%	99.86%	99.10%	99.67%	99.27%
Target [b]	**99.00%**	**99.01%**	**99.10%**	**99.50%**	**99.00%**
Margin [a-b]	**0.73%**	**0.85%**	**0.00%**	**0.17%**	**0.27%**

Logistic Breakdown Structure (LBS)

To analyse each component's RAM, the first step is to define the LBS (Logistic Breakdown Structure) of the system. LBS is the systematic decomposition of a system in physical terms. PBS (Product Breakdown Structure) is often used interchangeably with LBS. Since LBS is essential for techniques

such as FMECA and configuration identification, it should be developed in early phases and maintained carefully throughout the project.
Figure 33 shows an example of LBS for the rolling stock subcontractor. Other subcontractors should develop their own LBS for their deliverables, and all LBSs should be consolidated into a single document for the systems engineering team.
For practical use, it is beneficial for the LBS to be formatted as an Excel table, as illustrated in Figure 39. The lowest level of LBS should be the LRUs (Line Replaceable Units), considering that the maintenance personnel are expected to replace a failed part on-site with a spare part.

Figure 33 - Example of LBS of rolling stock

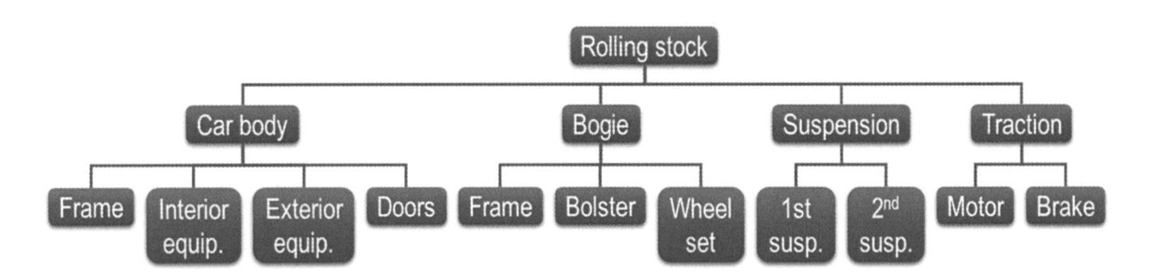

There are some principles in developing LBS as follows:

- The top and upper levels of LBS should be structured based on the contractual framework, as the LBS will be used to perform RAM analysis based on the RAM data of each supplier's equipment.
- The middle level of the LBS should be developed by suppliers (sub-contractors) according to their subsystem structure, facilitating suppliers in managing RAM data relevant to their deliverables.
- The lowest level of the LBS should be LRUs (Line Replaceable Units), based on the assumption that maintainers will replace components with spares when they fail or when the durability period expires.

RBD & FTA

< RBD >

RBD (Reliability Block Diagram) is a diagrammatic method used for illustrating how the reliability of component or subsystem contributes to failures of the overall system. RBD presents the relationships between failures within a system using logical symbols, as shown in Figure 34. However, the RBD may appear in different formats in technical documents.

It is used in RAM analysis to model the reliability and performance of a system. Originally, RBD was a method for depicting each component and its reliability in a diagram, illustrating the configuration between all components. However, RBD is currently employed as an analytical method rather than merely an expression method.

Each box depicted in RBD represents a module corresponding to a component or subsystem and can be used for explaining other failures that may affect system reliability. To conduct RAM predication and analysis, an RBD should contain at least the following information:

- Component ID
- Numeric data of MTBF (or failure rate) and MTTR
- Configuration type (serial or parallel)

Figure 34 - Example of RBD

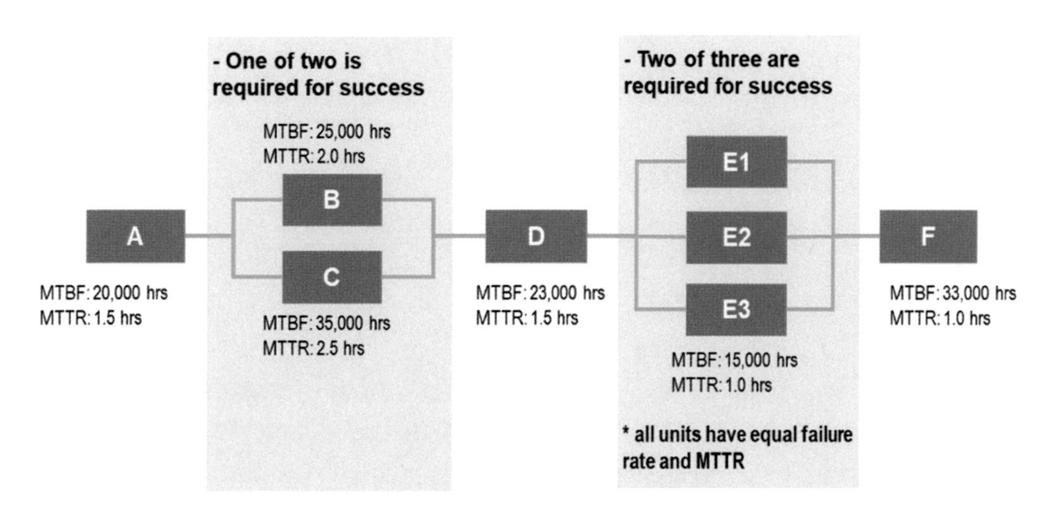

The configuration types include serial, parallel, and k-out-of-n. Each block in the diagram represents a component or subsystem and its configuration. This approach helps to:

- Identify weak points: Highlight critical components whose failure could lead to system downtime.
- Quantify reliability: Calculate the probability of system success or failure based on the reliability of individual components.
- Optimise design: Assess different types of configurations to improve system reliability and availability.

The process of RBD analysis is as follows:

- Understand the system thoroughly.
- Construct the RBD.
- Assign reliability and maintainability data to the components.
- Conduct quantitative analysis.

< FTA >

Fault Tree Analysis (FTA) is a systematic method used in engineering and risk management to identify and analyse potential failures within a system. It aids in understanding the causes of system failures or accidents by breaking down the failure into its component parts and tracing the pathways that lead to the failure event. It is used for calculating the overall MTBF (or failure rate) and MTTR of the entire system based on elements of the RBD.

The process of FTA is as follows:

- Identifying the top event: The analysis begins by identifying the undesired event or failures that need to be analysed, referred to as the "top event." This could be an accident, a system failure, or any event that needs to be prevented or mitigated.
- Identifying contributing events: Next, all possible factors or events that could contribute to the top event are identified. These are known as "basic events." Basic events typically include specific component failures, human errors, environmental conditions, or other factors that directly or indirectly lead to the top event. To calculate overall reliability or availability, the FTA should focus solely on component failures.
- Building the fault tree: Once basic events are identified, they are organised into a tree-like structure with the top event at the apex and the basic events at the base. Logical relationships between events are represented using logic gates such as AND, OR, and NOT gates (See Figure 35).
- Analysing the fault tree: The fault tree is analysed to assess the probability of the top event occurring based on the probabilities of the basic events and their logical relationships. Quantitative or qualitative analysis techniques can be employed, depending on the availability of data and the complexity of the system.
- Identifying mitigation measures: After completing the fault tree analysis, mitigation measures can be developed to enhance the overall reliability of the system.

Figure 35 – FTA symbols

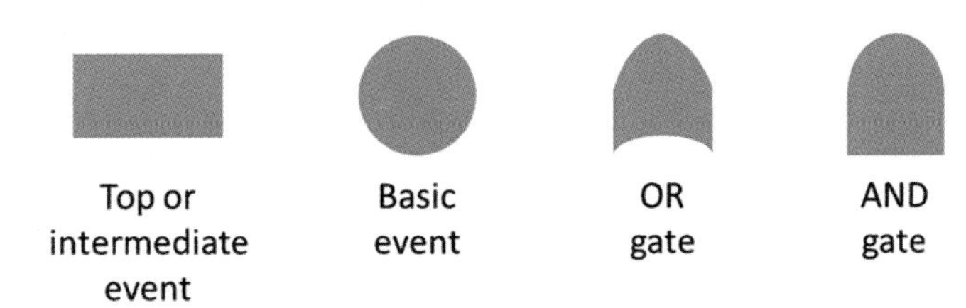

If the number of basic events is manageable, it is straightforward to calculate the overall reliability of the system manually. However, specialised FTA tools are necessary when the system contains lots of basic events.

Reliability optimisation

Reliability can be enhanced by decreasing the failure rate. Several methodologies to achieve this include:

- Reducing the number of subsystems and/or components.
- Utilising subsystems and/or components with high reliability.
- Adopting redundancy configurations within systems to minimise system failures.

Among the methodologies mentioned above, adding redundancy is frequently utilised, though it always incurs additional costs. Therefore, reliability should be optimised. Here is an example of reliability optimisation:

< Original >

Imagine that a system is required to meet a reliability target of 80%, and it consists of four components in a series configuration, as illustrated in Figure 36. The system's current reliability is 42.4% (= 85% x 70% x 75% x 95%), and it costs 22 USD.

Figure 36 - Original reliabilities

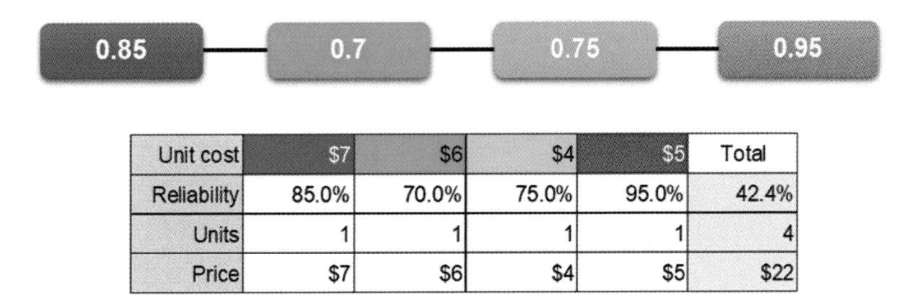

Unit cost	$7	$6	$4	$5	Total
Reliability	85.0%	70.0%	75.0%	95.0%	42.4%
Units	1	1	1	1	4
Price	$7	$6	$4	$5	$22

< Option 1 >

By adding redundant components to the sky-blue, green, and red ones, the system achieves a reliability of 81.2%, satisfying the required target, as illustrated in Figure 37. The total cost for this redundancy is 47 USD. In the case of green components in a parallel configuration, the reliability is calculated as follows:

$$1 - (1 - 0.75)^{\wedge}3 = 0.984$$

Figure 37 – Reliabilities of option 1

0.85 | 0.7 | 0.7 | 0.7 | 0.75 | 0.75 | 0.75 | 0.95 | 0.95

Unit cost	$7	$6	$4	$5	Total
Reliability	85.0%	97.3%	98.4%	99.8%	81.2%
Units	1	3	3	2	9
Price	$7	$18	$12	$10	$47

< Option 2 >

However, there is an alternative way to increase the reliability while reducing costs, as shown in Figure 38:

Figure 38 - Reliabilities of option 2

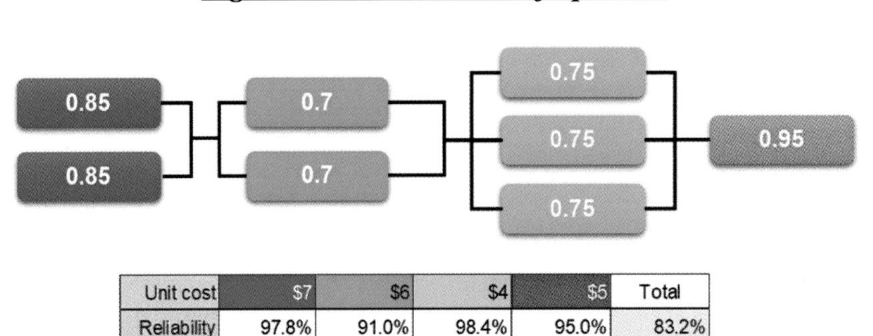

Unit cost	$7	$6	$4	$5	Total
Reliability	97.8%	91.0%	98.4%	95.0%	83.2%
Units	2	2	3	1	8
Price	$14	$12	$12	$5	$43

Comparing with Option 1, this option can improve reliability by an additional 2% and save 4 USD simply by adjusting the component composition. This approach is part of LCC (Life Cycle Cost) techniques for reliability optimisation.

Maintainability enhancement

Maintainability can be enhanced by reducing MTTR. Several methodologies to achieve this include:

- Simplifying the design to facilitate inspection by maintainers when failures occur.
- Adopting a higher level of LRU (Line Replaceable Unit).
- Identifying all test points and fault indicators.
- Ensuring accessibility to all components of system.
- Making disassembly and assembly straightforward.
- Standardising the design to help maintainers intuitively and easily understand the system schema.

Enhancing maintainability involves optimising the ease, speed, and cost-effectiveness of system maintenance. Here are some concrete strategies to improve maintainability:

- Modular design: Develop systems with modular components, allowing for the replacement or repair of individual parts without affecting the entire system. This reduces downtime and simplifies maintenance tasks.
- Standardisation of components: Use standardised and interchangeable parts across systems. This minimises the need for specialised tools or training and speeds up the replacement or repair process.
- Automated diagnostic tools: Implement automated monitoring and diagnostic tools that detect potential failures before they occur. Predictive maintenance can identify issues early, reducing the time and complexity of repairs.
- Clear and comprehensive documentation: Provide detailed maintenance manuals, schematics, and operational guides. This ensures that maintenance personnel have the necessary information to perform repairs accurately and efficiently.
- Training programs: Offer continuous training for maintenance personnel, ensuring they stay updated on the latest techniques, tools, and system upgrades. Skilled technicians can quickly diagnose and resolve issues, reducing overall maintenance time.
- Remote monitoring and control: Implement systems that allow remote monitoring and control of equipment. This enables technicians to diagnose and sometimes even fix problems remotely, reducing the need for on-site repairs.

RAM prediction

RAM prediction involves assessing and estimating the reliability, availability, and maintainability of a system or component. It typically employs modelling and analysis techniques to forecast how the system components will achieve the overall RAM target, particularly the overall availability. To analyse the overall availability, RAM engineers at the top level in the organisational structure of the project should prepare a worksheet which include LBS (Logistic Breakdown Structure), or PBS (Product Breakdown Structure) based on contractual agreement. The RAM engineer will gather all RAM-related data provided by system suppliers and input it into the worksheet.

With the overall MTBF and MTTR, the overall availability can be calculated using the formula below:

$$\text{Availability}_{\text{overall}}(\%) = \frac{\text{MTBF}_{\text{overall}}}{\text{MTBF}_{\text{overall}} + \text{MTTR}_{\text{overall}}} \times 100$$

Prediction process

The steps to perform RAM prediction are as follows:

- Gather RAM data relate to the railway service.
- If there are MCBF (Mean Cycle Between Failures), MDBF (Mean Distance Between Failures), or MKBF (Mean Kilometres Between Failures), convert them into MTBF (Mean Time Between Failures).
- Convert MTBFs into failure rates.
- Calculate the total failure rate of multiple units forming parallel configurations and “k out of n” configurations by converting the MTBF of component groups into that of a single configuration. All configurations can then be treated as a series configuration.
- Calculate the overall failure rate of all units.
- Inverse MTTRs of multiple components forming parallel configurations and “k out of n” configurations.
- Calculate the total MTTR of multiple units.
- Calculate the overall MTTR by calculating all failure rates and MTTRs.
- Calculate the overall availability with the overall MTBF and MTTR.

Here are steps to calculate the overall availability, along with an example of RAM data sheet (See Figure 39).

Figure 39 - RAM data

L1	L2	L3	LRU	Configuration type	MTBF (hr)	MCBF	MTTR (hr)
A	Signalling						
	A-1	Balise					
		A-1-1	**Balise 1**	**Serial**	**1,000**		**1**
		A-1-2	**Balise 2**	**Serial**	**4,000**		**2**
		A-1-3	**Balise 3**	**Serial**	**2,000**		**1**
	A-2	EIS (all on-line)					
		A-2-1	**EIS 1**	**Parallel (with repair)**	**200**		**5**
		A-2-2	**EIS 2**	**Parallel (with repair)**	**100**		**3**
B	Rolling stock						
	B-1	Braking system					
		B-1-1	**BS 1**	**Serial**	**3,000**		**2**
		B-1-2	**BS 2**	**Serial**	**1,000**		**4**
	B-2	Motor (without repair)					
		B-2-1	**Motor 1**	**2 out of 3 (without repair)**	**500**		**2**
		B-2-2	**Motor 2**	**2 out of 3 (without repair)**	**500**		**2**
		B-2-3	**Motor 3**	**2 out of 3 (without repair)**	**500**		**2**
C	AFC						
	C-1	AFC (entrance)					
		C-1-1	**AFC 1**	**Parallel (with repair)**		**10,000**	**5**
		C-1-2	**AFC 2**	**Parallel (with repair)**		**10,000**	**5**
	C-2	AFC (exit)					
		C-2-1	**AFC 3**	**Parallel (with repair)**		**10,000**	**5**
		C-2-2	**AFC 4**	**Parallel (with repair)**		**10,000**	**5**

Convert into MTBF and failure rate

< MTBF >

There can be non-time-based data such as MCBF or MKBF, addressed in the chapter "MKBF and MCBF". These non-time-based data should be converted into MTBF because only time-based data can be used in calculating the overall availability.

AFCs (Automated Fare Collectors) have MCBF data inFigure 40, which should be converted into MTBF data by following the steps outlined in the chapter "MKBF and MCBF". As a result of the calculation, each AFC for boarding has an MTBF of 100 hours, while the AFC for alighting has an MTBF of 200 hours (see "MTBF' (hr)" column of the table), as explained in the chapter.

Figure 40 - Converted into MTBF

L1	L2	L3	LRU	Configuration type	MTBF (hr)	MCBF	MTTR (hr)	MTBF' (hr)
A	Signalling							
	A-1	Balise						
		A-1-1	Balise 1	Serial	1,000		1	1,000
		A-1-2	Balise 2	Serial	4,000		2	4,000
		A-1-3	Balise 3	Serial	2,000		1	2,000
	A-2	EIS (all on-line)						
		A-2-1	EIS 1	Parallel (with repair)	200		5	200
		A-2-2	EIS 2	Parallel (with repair)	100		3	100
B	Rolling stock							
	B-1	Braking system						
		B-1-1	BS 1	Serial	3,000		2	3,000
		B-1-2	BS 2	Serial	1,000		4	1,000
	B-2	Motor (without repair)						
		B-2-1	Motor 1	2 out of 3 (without repair)	500		2	500
		B-2-2	Motor 2	2 out of 3 (without repair)	500		2	500
		B-2-3	Motor 3	2 out of 3 (without repair)	500		2	500
C	AFC							
	C-1	AFC (entrance)						
		C-1-1	AFC 1	Parallel (with repair)		10,000	5	100
		C-1-2	AFC 2	Parallel (with repair)		10,000	5	100
	C-2	AFC (exit)						
		C-2-1	AFC 3	Parallel (with repair)		10,000	5	200
		C-2-2	AFC 4	Parallel (with repair)		10,000	5	200

< Failure rates >

A certain subsystem may consist of many redundancies grouped to perform a specific function, which is regarded as a single component. This configuration is referred to as a "parallel configuration," and the overall failure rate of the group needs to be calculated. The purpose of this step is to treat a group of parallel components as a single component. This overall failure rate will also be used in FMECA (Failure Mode, Effect, and Criticality Analysis) as the overall failure frequency of the grouped components.

The overall MTBF of a component group will be calculated based on their configuration types, such as parallel and "k out of n." To calculate the overall MTBF of the groups (A-2, B-2, C-1, and C-2), each MTBF should be first converted into failure rates, as shown in Figure 41 (Column "F.r. (hr)").

Figure 41 - Converted into failure rate

L1	L2	L3	LRU	Configuration type	MTBF (hr)	MCBF	MTTR (hr)	MTBF' (hr)	F.r. (hr)
A	Signalling								
	A-1	Balise							
		A-1-1	Balise 1	Serial	1,000		1	1,000	0.001000
		A-1-2	Balise 2	Serial	4,000		2	4,000	0.000250
		A-1-3	Balise 3	Serial	2,000		1	2,000	0.000500
	A-2	EIS (all on-line)							
		A-2-1	EIS 1	Parallel (with repair)	200		5	200	0.005000
		A-2-2	EIS 2	Parallel (with repair)	100		3	100	0.010000
B	Rolling stock								
	B-1	Braking system							
		B-1-1	BS 1	Serial	3,000		2	3,000	0.000333
		B-1-2	BS 2	Serial	1,000		4	1,000	0.001000
	B-2	Motor (without repair)							
		B-2-1	Motor 1	2 out of 3 (without repair)	500		2	500	0.002000
		B-2-2	Motor 2	2 out of 3 (without repair)	500		2	500	0.002000
		B-2-3	Motor 3	2 out of 3 (without repair)	500		2	500	0.002000
C	AFC								
	C-1	AFC (entrance)							
		C-1-1	AFC 1	Parallel (with repair)		10,000	5	100	0.010000
		C-1-2	AFC 2	Parallel (with repair)		10,000	5	100	0.010000
	C-2	AFC (exit)							
		C-2-1	AFC 3	Parallel (with repair)		10,000	5	200	0.005000
		C-2-2	AFC 4	Parallel (with repair)		10,000	5	200	0.005000

As explained in the earlier chapter, the formula is as follows:

$$\text{Failure rate}_{\text{each}}\ (\text{hr}) = \frac{1}{\text{MTBF}_{\text{each}}}$$

Calculations

< Failure rate of multiple configurations >

When it comes to the motors, at least two out of three motors are required to operate for success. The overall failure rate can be calculated as explained in section on "Failure rate." The column "F.R.' (hr)" on the far left in Figure 42 shows the overall failure rates of the grouped units configured in parallel and "k out of n" (A-2, B-2, C-1, and C-2). Please refer to the chapter on "Failure rate calculation" for details on how to calculate parallel and "k out of n" configurations.

Figure 42 - Overall failure rates of multiple units

L1	L2	L3	LRU	Configuration type	MTBF (hr)	MCBF	MTTR (hr)	MTBF' (hr)	F.r. (hr)	F.R.' (hr)
A	Signalling									
	A-1	Balise								
		A-1-1	Balise 1	Serial	1,000		1	1,000	0.001000	0.00100000
		A-1-2	Balise 2	Serial	4,000		2	4,000	0.000250	0.00025000
		A-1-3	Balise 3	Serial	2,000		1	2,000	0.000500	0.00050000
	A-2	EIS (all on-line)								0.00022264
		A-2-1	EIS 1	Parallel (with repair)	200		5	200	0.005000	
		A-2-2	EIS 2	Parallel (with repair)	100		3	100	0.010000	
B	Rolling stock									
	B-1	Braking system								
		B-1-1	BS 1	Serial	3,000		2	3,000	0.000333	0.00033333
		B-1-2	BS 2	Serial	1,000		4	1,000	0.001000	0.00100000
	B-2	Motor (without repair)								0.00240000
		B-2-1	Motor 1	2 out of 3 (without repair)	500		2	500	0.002000	
		B-2-2	Motor 2	2 out of 3 (without repair)	500		2	500	0.002000	
		B-2-3	Motor 3	2 out of 3 (without repair)	500		2	500	0.002000	
C	AFC									
	C-1	AFC (entrance)								0.00004000
		C-1-1	AFC 1	Parallel (with repair)		10,000	5	100	0.010000	
		C-1-2	AFC 2	Parallel (with repair)		10,000	5	100	0.010000	
	C-2	AFC (exit)								0.00001000
		C-2-1	AFC 3	Parallel (with repair)		10,000	5	200	0.005000	
		C-2-2	AFC 4	Parallel (with repair)		10,000	5	200	0.005000	

< Overall failure rate >

The overall failure rate for all units should be calculated as shown in Figure 43. The overall failure rate of the system is 0.005755.

Figure 43 - Total failure rate

L1	L2	L3	LRU	Configuration type	MTBF (hr)	MCBF	MTTR (hr)	MTBF' (hr)	F.r. (hr)	F.R.' (hr)
A	Signalling									
	A-1	Balise								
		A-1-1	Balise 1	Serial	1,000		1	1,000	0.001000	0.00100000
		A-1-2	Balise 2	Serial	4,000		2	4,000	0.000250	0.00025000
		A-1-3	Balise 3	Serial	2,000		1	2,000	0.000500	0.00050000
	A-2	EIS (all on-line)								0.00022264
		A-2-1	EIS 1	Parallel (with repair)	200		5	200	0.005000	
		A-2-2	EIS 2	Parallel (with repair)	100		3	100	0.010000	
B	Rolling stock									
	B-1	Braking system								
		B-1-1	BS 1	Serial	3,000		2	3,000	0.000333	0.00033333
		B-1-2	BS 2	Serial	1,000		4	1,000	0.001000	0.00100000
	B-2	Motor (without repair)								0.00240000
		B-2-1	Motor 1	2 out of 3 (without repair)	500		2	500	0.002000	
		B-2-2	Motor 2	2 out of 3 (without repair)	500		2	500	0.002000	
		B-2-3	Motor 3	2 out of 3 (without repair)	500		2	500	0.002000	
C	AFC									
	C-1	AFC (entrance)								0.00004000
		C-1-1	AFC 1	Parallel (with repair)		10,000	5	100	0.010000	
		C-1-2	AFC 2	Parallel (with repair)		10,000	5	100	0.010000	
	C-2	AFC (exit)								0.00001000
		C-2-1	AFC 3	Parallel (with repair)		10,000	5	200	0.005000	
		C-2-2	AFC 4	Parallel (with repair)		10,000	5	200	0.005000	
Total										0.00575597

The formula to calculate the overall failure rate of a series configuration is as follows:

$$\frac{1}{\mathit{Overall}\ \mathrm{MTBF}_{\mathrm{series}}} = \mathrm{F.r}_1 + \mathrm{F.r}_2 + \mathrm{F.r}_3 + \cdots + \mathrm{F.r}_n$$

You might think that the equation is not for the failure rate, but please remember that failure rate and MTBF are reciprocal.

$$\text{Failure rates (hr)} = \frac{1}{\mathrm{MTBF}}$$

< Inverse of MTTRs >

Since MTTR data or MTTR-related data are always provided as numbers, there is no need to convert it. You only need to calculate the sum of components' MTTR according to their configuration type such as series, parallel, or "k out of n" configuration.

To calculate the overall MTTRs of parallel and "k out of n" configurations, the inverse of MTTRs should be calculated, as shown in Figure 44.

Figure 44 - Inverse of MTTRs

L1	L2	L3	LRU	Configuration type	MTBF (hr)	MTTR (hr)	F.R.' (hr)	1 / MTTR (hr)
A	Signalling							
	A-1	Balise						
		A-1-1	**Balise 1**	**Serial**	**1,000**	**1**	**0.00100000**	
		A-1-2	**Balise 2**	**Serial**	**4,000**	**2**	**0.00025000**	
		A-1-3	**Balise 3**	**Serial**	**2,000**	**1**	**0.00050000**	
	A-2	EIS (all on-line)					0.00022264	0.53333
		A-2-1	**EIS 1**	**Parallel (with repair)**	**200**	**5**		**0.20000**
		A-2-2	**EIS 2**	**Parallel (with repair)**	**100**	**3**		**0.33333**
B	Rolling stock							
	B-1	Braking system						
		B-1-1	**BS 1**	**Serial**	**3,000**	**2**	**0.00033333**	
		B-1-2	**BS 2**	**Serial**	**1,000**	**4**	**0.00100000**	
	B-2	Motor (without repair)					0.00240000	-
		B-2-1	**Motor 1**	**2 out of 3 (without repair)**	**500**	**2**		
		B-2-2	**Motor 2**	**2 out of 3 (without repair)**	**500**	**2**		
		B-2-3	**Motor 3**	**2 out of 3 (without repair)**	**500**	**2**		
C	AFC							
	C-1	AFC (entrance)					0.00004000	0.40000
		C-1-1	**AFC 1**	**Parallel (with repair)**		**5**		**0.20000**
		C-1-2	**AFC 2**	**Parallel (with repair)**		**5**		**0.20000**
	C-2	AFC (exit)					0.00001000	0.40000
		C-2-1	**AFC 3**	**Parallel (with repair)**		**5**		**0.20000**
		C-2-2	**AFC 4**	**Parallel (with repair)**		**5**		**0.20000**

< All MTTRs >

All MTTRs of parallel and "k out of n" configurations should be calculated as shown in Figure 45.

Figure 45 - all MTTRs

L1	L2	L3	LRU	Configuration type	MTBF (hr)	MTTR (hr)	F.R.' (hr)	1 / MTTR (hr)	MTTR' (hr)
A	Signalling								
	A-1	Balise							
		A-1-1	**Balise 1**	**Serial**	**1,000**	**1**	**0.00100000**		**1.00**
		A-1-2	**Balise 2**	**Serial**	**4,000**	**2**	**0.00025000**		**2.00**
		A-1-3	**Balise 3**	**Serial**	**2,000**	**1**	**0.00050000**		**1.00**
	A-2	EIS (all on-line)					0.00022264	0.53333	1.88
		A-2-1	**EIS 1**	**Parallel (with repair)**	**200**	**5**		**0.20000**	
		A-2-2	**EIS 2**	**Parallel (with repair)**	**100**	**3**		**0.33333**	
B	Rolling stock								
	B-1	Braking system							
		B-1-1	**BS 1**	**Serial**	**3,000**	**2**	**0.00033333**		**2.00**
		B-1-2	**BS 2**	**Serial**	**1,000**	**4**	**0.00100000**		**4.00**
	B-2	Motor (without repair)					0.00240000	-	1.00
		B-2-1	**Motor 1**	**2 out of 3 (without repair)**	**500**	**2**			
		B-2-2	**Motor 2**	**2 out of 3 (without repair)**	**500**	**2**			
		B-2-3	**Motor 3**	**2 out of 3 (without repair)**	**500**	**2**			
C	AFC								
	C-1	AFC (entrance)					0.00004000	0.40000	2.50
		C-1-1	**AFC 1**	**Parallel (with repair)**		**5**		**0.20000**	
		C-1-2	**AFC 2**	**Parallel (with repair)**		**5**		**0.20000**	
	C-2	AFC (exit)					0.00001000	0.40000	2.50
		C-2-1	**AFC 3**	**Parallel (with repair)**		**5**		**0.20000**	
		C-2-2	**AFC 4**	**Parallel (with repair)**		**5**		**0.20000**	

For parallel configuration, the following formula is used to calculate the overall MTTR of a group (you can obtain the overall MTTR by inversing the sum of them). The equation to calculate the overall MTTR of parallel configuration is:

$$\frac{1}{\mathit{Overall}\ \mathrm{MTTR_{parallel}}} = \frac{1}{\mathrm{MTTR_1}} + \frac{1}{\mathrm{MTTR_2}} + \cdots + \frac{1}{\mathrm{MTTR_n}}$$

The formula for calculating the overall MTTR of "k out of n" configuration is straightforward, as shown below. Therefore, component B-2 in the data sheet does not need to have reciprocal value ("1/MTTR (hr)").

$$\mathit{Overall}\ \mathrm{MTTR_{(k\ out\ of\ n)}} = \frac{\mathrm{MTTR}}{\mathrm{n - k + 1}}$$

These overall MTTRs can be used in FMECA (Failure Mode, Effect, and Criticality Analysis) as the overall severity of grouped components if it is determined that severity is calculated using the MTTRs.

< MTTR multiplied by failure rates >

The column "Fr*MTTR" on the far left in Figure 46 shows the product of MTTRs and failure rates, which are used in the calculation of the overall MTTR, as explained in the section on "Maintainability."

Figure 46 – Overall MTTR multiplied by failure rates and MTTR

L1	L2	L3	LRU	Configuration type	MTBF (hr)	MTTR (hr)	F.R.' (hr)	1 / MTTR (hr)	MTTR' (hr)	Fr*MTTR
A	Signalling									
	A-1	Balise								
		A-1-1	Balise 1	Serial	1,000	1	0.00100000		1.00	0.001000
		A-1-2	Balise 2	Serial	4,000	2	0.00025000		2.00	0.000500
		A-1-3	Balise 3	Serial	2,000	1	0.00050000		1.00	0.000500
	A-2	EIS (all on-line)					0.00022264	0.53333	1.88	0.000417
		A-2-1	EIS 1	Parallel (with repair)	200	5		0.20000		
		A-2-2	EIS 2	Parallel (with repair)	100	3		0.33333		
B	Rolling stock									
	B-1	Braking system								
		B-1-1	BS 1	Serial	3,000	2	0.00033333		2.00	0.000667
		B-1-2	BS 2	Serial	1,000	4	0.00100000		4.00	0.004000
	B-2	Motor (without repair)					0.00240000	-	1.00	0.002400
		B-2-1	Motor 1	2 out of 3 (without repair)	500	2				
		B-2-2	Motor 2	2 out of 3 (without repair)	500	2				
		B-2-3	Motor 3	2 out of 3 (without repair)	500	2				
C	AFC									
	C-1	AFC (entrance)					0.00004000	0.40000	2.50	0.000100
		C-1-1	AFC 1	Parallel (with repair)		5		0.20000		
		C-1-2	AFC 2	Parallel (with repair)		5		0.20000		
	C-2	AFC (exit)					0.00001000	0.40000	2.50	0.000025
		C-2-1	AFC 3	Parallel (with repair)		5		0.20000		
		C-2-2	AFC 4	Parallel (with repair)		5		0.20000		
			Total				0.00575597			0.009609

The overall failure rate of all units is 0.00575597, and the overall value of the MTTRs multiplied by failure rates is 0.009609, respectively.

Overall MTBF, MTTR and availability

< Overall MTBF >

The overall MTBF is easily calculated by inversing the overall failure rate. The overall MTBF is 173.733 hours (= 1 / 0.00575597).

< Overall MTTR >

The overall MTBF is 1.669 hours, calculated as follows:

$$1.669 = \frac{0.009609}{0.00575597}$$

$$\text{Overall MTTR}_{\text{series}} = \frac{\text{F.r}_1 \times \text{MTTR}_1 + \text{F.r}_2 \times \text{MTTR}_2 + \cdots + \text{F.r}_n \times \text{MTTR}_n}{\text{F.r}_1 + \text{F.r}_2 + \cdots + \text{F.r}_n}$$

< Overall availability >

The overall MTBF is 173.733 hours, and the overall MTTR is 1.669 hours; thus, the overall availability of the system is 0.99048.

$$0.99048 = \frac{0.009609}{0.009609 + 0.00575597}$$

$$\text{Availability}_{\text{overall}} = \frac{\text{MTBF}_{\text{overall}}}{\text{MTBF}_{\text{overall}} + \text{MTTR}_{\text{overall}}}$$

Calculation of total availability of different groups

If there are two systems managed separately by two different teams, as shown in Figure 47, the overall availability can be calculated in two ways:

- By treating the two systems as components.
- By multiplying the availabilities of the two systems together.

Figure 47 - MTBFs and MTTRs of two systems

System	MTBF	MTTR [a]	Failure rate [b]	Fr x MTTR [a x b]
System 1	a	b	1/a	b/a
System 2	c	d	1/c	d/c

< Treating them as components >

You can calculate the overall availability of the two systems by using the following simplified equations with an asterisk "*":

Overall MTBFs	Overall MTTRs
$\frac{1}{\mathrm{MTBF_{total}}} = \frac{1}{a} + \frac{1}{c}$	$\mathrm{MTTR_{total}} = \frac{\frac{b}{a} + \frac{d}{c}}{\frac{1}{a} + \frac{1}{c}}$
▼	▼
$\frac{1}{\mathrm{MTBF_{total}}} = \frac{a + c}{ac}$	$\mathrm{MTTR_{total}} = \frac{\frac{ad + bc}{ac}}{\frac{a + c}{ac}}$
▼	▼
* $\mathrm{MTBF_{total}} = \frac{ac}{a + c}$	* $\mathrm{MTTR_{total}} = \frac{ad + bc}{a + c}$

With the overall MTBF and MTTR, you can calculate the overall availability:

$$\mathrm{Availability_{total}} = \frac{\frac{ac}{a + c}}{\frac{ac}{a + c} + \frac{ad + bc}{a + c}}$$

▼

$$\mathrm{Availability_{total}} = \frac{\frac{ac}{a + c}}{\frac{ac + ad + bc}{a + c}}$$

▼

$$* \ \mathrm{Availability_{total}} = \frac{ac}{ac + ad + bc}$$

< Multiplying availabilities >

There is another formula commonly introduced on many websites for calculating the overall availability of several systems, as follows:

$$\text{Total A (hr)}_{\text{series}} = A_1 \times A_2 \times \cdots \times A_n$$

where:

- Total A $(hr)_{series}$ is the overall availabilities of the systems.
- A_n is the overall availability of each system.

Let's compare both formulae. Figure 48 represents the calculation using each system's MTBF and MTTR explained earlier, while Figure 49 uses the availability calculated by the formula mentioned above. The overall availabilities differ from each other: Figure 48 shows 0.999108, whereas Figure 49 shows 0.99106.

Figure 48 - Calculation by MTBFs and MTTRs

	MTBF	MTTR	Failure rate	Fr x MTTR	Availability
Group 1	5,000	25	0.0002	0.005	
Group 2	10,000	40	0.0001	0.004	
all	3,333.3	30			0.991080278

Figure 49 - Calculation by availabilities

	MTBF	MTTR	Failure rate	Fr x MTTR	Availability
Group 1	5,000	25	0.0002	0.005	0.995024876
Group 2	10,000	40	0.0001	0.004	0.996015936
all					0.991060633

Considering Figure 47, the equation for Figure 49 is as follows:

$$\text{Availability}_{\text{total}} = \frac{a}{a+b} \times \frac{c}{c+d}$$

▼

$$^{*}\text{Availability}_{\text{total}} = \frac{ac}{ac+ad+bc+bd}$$

Comparing both equations, we can find that there is a difference between them, especially:

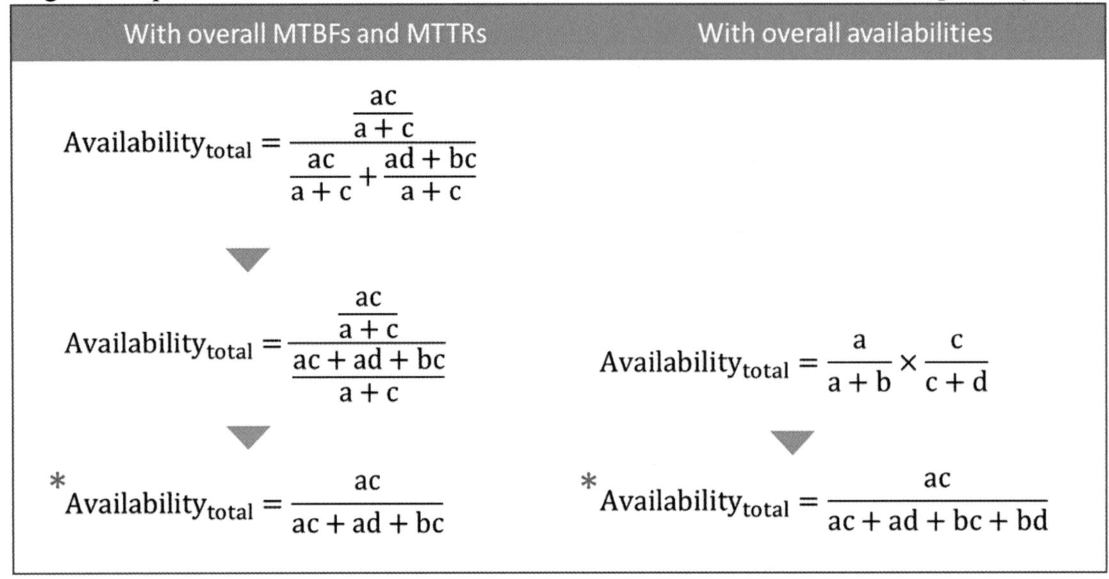

With overall MTBFs and MTTRs	With overall availabilities
$\text{Availability}_{\text{total}} = \dfrac{\frac{ac}{a+c}}{\frac{ac}{a+c} + \frac{ad+bc}{a+c}}$	
$\text{Availability}_{\text{total}} = \dfrac{\frac{ac}{a+c}}{\frac{ac+ad+bc}{a+c}}$	$\text{Availability}_{\text{total}} = \dfrac{a}{a+b} \times \dfrac{c}{c+d}$
* $\text{Availability}_{\text{total}} = \dfrac{ac}{ac+ad+bc}$	* $\text{Availability}_{\text{total}} = \dfrac{ac}{ac+ad+bc+bd}$

FMECA (Failure Mode, Effect and Criticality Analysis)

FMECA is widely used in RAM analysis and safety analysis, maintainability analysis, logistic support by identifying and quantifying sources of undesirable failure modes. It serves to verify design integrity and document RAM and safety risks. As the name implies, "Failure Mode and Effect Analysis" (FMEA) pertains to hazard analysis, while "Criticality Analysis" refers to risk evaluation.

- FMECA = FMEA + Criticality Analysis

Based on criticality analysis, each potential failure can be ranked according to the combination of severity influence and occurrence probability. The purposes of conducting FMECA are as follows:

- To identify single points of failures (SPOF) and understand requirements for redundancy.
- To provide an auditable method for identifying equipment failure modes and consequences.
- To ensure that the RAM engineer has a thorough understanding of the operation of a system under faulty conditions.
- To avoid the need for costly modifications in service by identifying design deficiencies early in the design phase.
- To assist in formulating maintenance strategies and serve as the basis for the failure diagnosis.

Required information

FMECA may employ two approaches for analysis: functional approach and hardware approach. The required information to perform the analysis is as follows:

- System definition (including system boundaries and expected operating environment) to develop the functional analysis worksheets
- System requirements specification to develop functional block diagram
- System architecture description

Each component in a system is rigorously assessed in FMECA activity in terms of the following:

- Modes of failure
- Causes
- Effects
- Detection methods and mitigation measures
- Criticality

For criticality analysis, the following standards are commonly used:

- Mil-Std-1629A: This standard is used to develop criticality analysis in terms of reliability and safety. It also requires the creation of Safety Criticality Items List (SCIL) and Reliability Criticality Items List (RCIL)
- BS EN 60812: This standard links criticality analysis with risk analysis and employs risk matrices or risk priority numbers (RPN)

Figure 50 shows an overview of the FMECA activities.

Figure 50 – FMECA activities

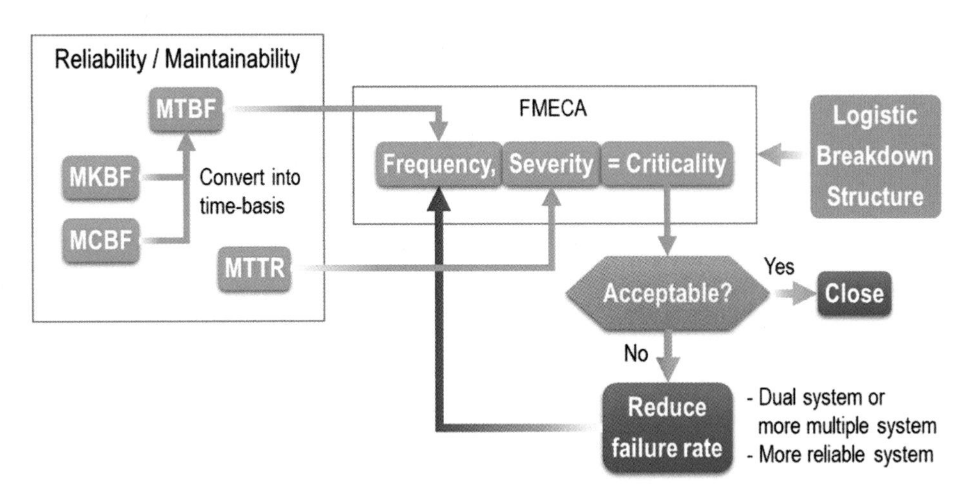

- At first, the format of the FMECA worksheet is prepared with the LBS provided by subsystem contractors.
- MTBF and MTTR data should be provided by suppliers. If the suppliers have provided MKBF (Mean Kilometres Between Failures) or MCBF (Mean Cycle Between Failures) instead of MTBF, MKBF and MCBF should be converted into MTBF.
- By analysing the criticality based on the risk matrix, the systems engineer can determine whether a failure (or criticality, or risk) is acceptable. If a failure is deemed unacceptable, the relevant system configuration should be modified by adding redundant components, resulting in a redundancy design to decrease the overall failure rate of the grouped components. It is also acceptable to replace the unit with another that has a significantly lower MTTR and/or a higher MTBF.

FMECA process

FMECA processes are used to identify potential system failure modes, their causes, and effects on system performance. It employs inductive techniques to examine every component in a system by asking the following questions:

- How can a component fail?
- What is the impact of the failure when the component fails – locally and system-wide?

Figure 51 – FMECA process

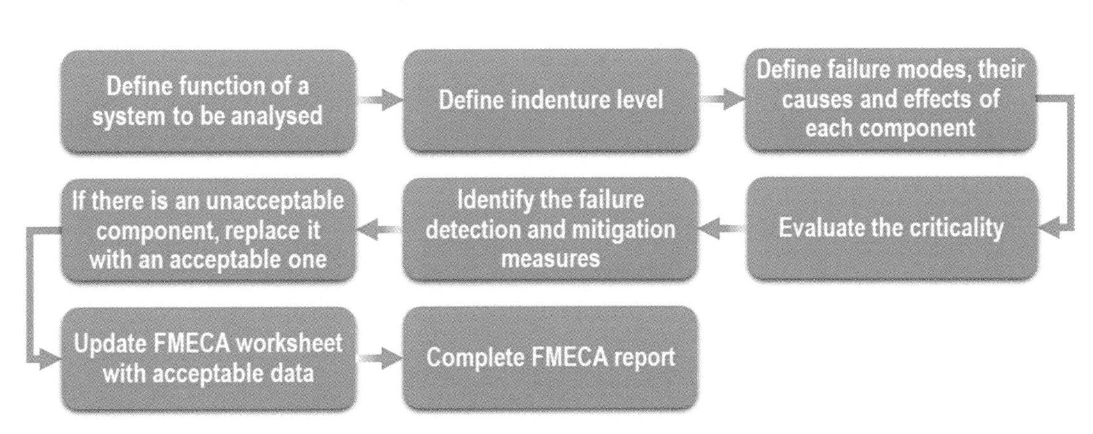

Figure 51 shows an overview of the FMECA process.

- First, functions of the target system should be identified to define the system boundaries.
- The indenture level indicates the hierarchy of the system components, which is important as it should reflect the RAM apportionment.
- Defining failure modes and their causes of each component is a crucial step in engineers' failure analysis. Defining effects (or severity) represents how a system engineer intends to manage the system safety, as the effect becomes the managerial target.
- Criticality can be easily calculated using the risk matrix, which will be introduced in another chapter.
- If there are effective mitigation measure for a failure, the frequency or severity of the failure can be reduced.
- If a component is deemed unacceptable due to the criticality analysis based on the risk matrix, the frequency and/or severity should be reduced until the criticality is acceptable.

< Step 1: Definition of system functions >

In this step, three things need to be defined as follows:

- Function of systems under different operating conditions: Clearly delineate how the system performs and operates across various environmental and operational scenarios to ensure a comprehensive evaluation of its roles and capabilities.
- System boundaries: Establish precise physical and functional boundaries of the system, defining the scope and interaction with external elements, which is crucial for accurate system analysis and design.
- Ground rules of system failure: Formulate fundamental principles and procedures for system failure scenarios to maintain consistency in fault analysis and recovery processes, thereby enhancing the system's overall reliability.

It is beneficial to develop FBD (Functional Block Diagram) for the analysis of system functions. An FBD is a logical diagram that illustrates the relationships between input variables and output variables. Figure 52 represents an example of a simplified FBD of rolling stock in terms of power supply, which should have more detailed input and output variables. Each function is represented as a set of elementary blocks, with input and output variables connected to these blocks by connection lines. The benefits of using an FBD include the following:

- Illustration of operation: It shows the operation, interrelationships, interdependencies, and functional entities of systems under various operating conditions.
- Tracing failure mode effects: It provides the ability to trace failure mode effects through all levels of indenture.
- Display of alternative modes of operation: It visually presents alternative modes of operation.
- Functional flow sequence: If offers a functional flow sequence for the system.

Figure 52 - Example of simplified FBD of rolling stock

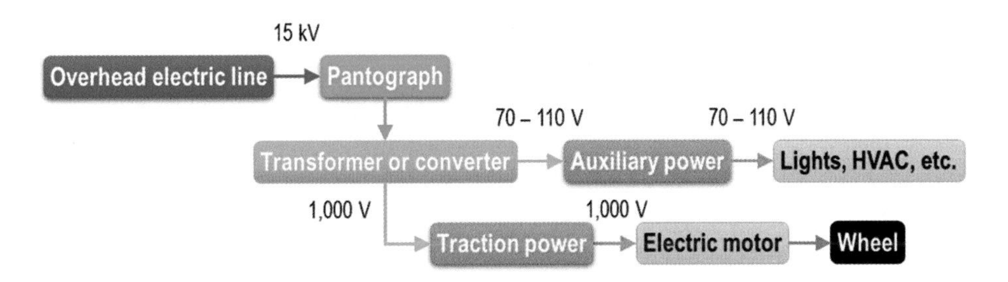

< Step 2: Indenture table >

An indenture table is a hierarchical system breakdown structure that outlines the relationships among the system, subsystems, equipment, components or LRUs (Line Replaceable Units). Before developing the indenture table, a numbering system should be established to identify each component or LRU for analysis. The indenture table should contain at least the following information:

- Reference ID: A unique identifier for each component or LRU.
- Subsystem (or assembly): The specific subsystem or assembly to which the component belongs.
- Location (if applicable): The physical location of the component within the system.
- Component: The name or designation of the component or LRU.
- Function: A brief description of the function performed by the component.

< Step 3: Identification of component failure modes, causes and effects >

Since the definitions of failure modes, causes and effects are sometimes confused, it is essential to define them clearly:

- Failure modes: The status in which the required function cannot be performed.
- Causes: The factors or conditions that lead to failure modes.
- Effects: The consequence resulting from failure modes.

Failure modes can be derived from three approaches as follows:

- Functional approach: Treating a system as a "black box" that provides specific required functions.
- Hardware approach: Considering the single failure modes of each individual component within the system.
- Process approach: Identifying risks from various aspects, such as staff, methods, material, machinery, and the environment.

Failure causes can be derived as follows:

- Identify all potential causes associated with each possible failure mode and within the adjacent indenture levels.
- Recognise all potential independent causes.
- Uncover secondary effects and devise recommended corrective actions.
- Consider human errors and software errors.

When deriving failure effects, the following effects should be considered:

- Local effect
- Effect on next higher assembly level
- Overall system effect

< Step 4: Criticality analysis >

Risk matrix (or criticality matrix) is used to analyse the criticality in many cases, especially in FMECA, as shown in Figure 53. To define the level of criticality, a risk or criticality matrix should be prepared first, which is generally defined in the RAM and Safety Management Plans or Assurance Management Plan.

Figure 53 – Example of risk matrix

			Severity			
			4	3	2	1
			Insignificant	Marginal	Critical	Catastrophic
Frequency	A	Frequent	R3	R4	R4	R4
	B	Probable	R2	R3	R4	R4
	C	Occasional	R1	R2	R3	R4
	D	Remote	R1	R1	R2	R3
	E	Improbable	R1	R1	R1	R2
	F	Incredible	R1	R1	R1	R1

Performing risk analysis, each component is evaluated to determine whether it is acceptable, as shown in Figure 54. If a component is ranked “R4-rated,” it should be replaced with another one that has a lower failure frequency or low severity, or both.

Figure 54 - Example of risk ranking

Risk Ranking	Category	Action to be applied against each category
R4	**Intolerable**	**Shall be eliminated.**
R3	**Undesirable**	**Shall only be accepted when risk reduction is practicable and under the agreement of Client, as appropriate.**
R2	**Tolerable**	**Acceptable with adequate control and the agreement of Client**
R1	**Negligible**	**Acceptable without any agreement.**

Components ranked "R3" or "R2" should be classified as SCIs (Safety Critical Items), and plans for managing them should be prepared.

< Step 5: Failure detection methods and mitigation measure >

The purpose of this step is to lower the risk ranking of each component, as shown in Figure 55. The following activities are conducted in this step:

- Identify and evaluate the failure detection methods and mitigation controls for each identified failure mode.
- Consider these controls in terms of the frequency and severity classifications.
- Identify corrective design or other actions required to eliminate the failure or control the risks.
- Identify other system attributes, such as logistic support.

Figure 55 - Decrease of risk ranking

		Severity			
		Insignificant	Marginal	Critical	Catastrophic
Frequency	Frequent	R3	R4	R4	R4
	Probable	R2	R3	R4	R4
	Occasional	R2	R2	R3	R4
	Remote	R1	R2	R2	R3
	Improbable	R1	R1	R2	R2
	Incredible	R1	R1	R1	R2

< Step 6: Elimination of unacceptable components >

Even though mitigation measures were performed in the previous step, there may still be components ranked as "R4". Any components ranked as "R4" should not be used in the system until redundant components are added, resulting in a parallel configuration.

< Step 7: FMECA worksheet >

The FMECA worksheet should contain at least the following:

- PBS (Product Breakdown Structure): indenture level and ID number (if any). The lowest level of breakdown should be LRU (Line Replaceable Unit).
- Item: component name
- Function: description of the component's function
- Failure mode: how a failure of a component is observed
- Failure cause: specific causes of the failure mode
- Failure effect: specific effect to subsystem (local) and system level
- Failure detection: how the occurrence of a failure mode is detected by an operator
- Alternative failure detection: how the occurrence of a failure mode is detected by a maintainer
- Time affected (minutes): estimated recovery time
- Service effect (Y/N): whether the failure is related to service
- Safety effect (Y/N): whether the failure is related to safety
- SSHA ID: relevant SSHA ID if "Safety effect" is marked "Y"
- Unit failure: failure rate per operating hour
- Failure mode ratio: the fraction of the part failure rate related to the failure mode under consideration
- Modal failure rate: the product of the "Unit failure" and "Failure mode ratio"
- Data source: reference for the unit failure rate and modal apportionment used
- Action to restore system: description of actions that restore functionality following the failure mode
- Frequency class: original level of failure frequency before confirmation of the mitigation measures
- Severity class: original level of severity
- Risk rank: original risk rank
- Mitigation measure: description of mitigation measures
- Frequency level: level of frequency after confirmation of the mitigation measures
- Severity level: level of severity after confirming the mitigation measures

< Step 8: Development of FMEACA report >
The report should contain the following topics:

- A description of the system that was analysed
- The indenture levels of the system
- Adopted approach
- Details of any data sources used and justification for their selection
- Any functional block diagrams and their hierarchy breakdown produced for the analysis
- The result, including the completed worksheets
- Any recommendations for further analysis, design changes, or modifications to reduce risks.

SCIs (Safety Critical Items)

A SCI is a component that has been determined to potentially contribute to a catastrophic hazard. As a result of performing FMECA, some components may have high-risk ranking based on their failure rate and the impact of failure.

Figure 56 - Risk ranking

Risk Ranking	Category	Action to be applied against each category
R4	**Intolerable**	**Shall be eliminated.**
R3	**Undesirable**	**Shall only be accepted when risk reduction is practicable and under the agreement of Client, as appropriate.**
R2	**Tolerable**	**Acceptable with adequate control and the agreement of Client**
R1	**Negligible**	**Acceptable without any agreement.**

According to risk ranking, each item will be determined to be accepted or not (See Figure 56). Items rated R1 are acceptable, while those rated R4 are not. On the other hand, items rated R2 and R3 need to be managed carefully during the operation and maintenance phase, as they are classified as SCIs.
SCIs require stringent maintenance controls to ensure that they continue to satisfactorily perform the required safety functions. The methodologies of maintenance controls of SCIs are as follows:

- Ensure that staff carrying out maintenance works on SCIs are adequately trained, certified, and qualified.
- The maintenance team should determine the periodic cycle of re-certifying and re-qualifying staff based on the nature of the maintenance works.
- The maintenance team should regularly update the SCIs list.

RAM demonstration

RAM demonstration is the process of verifying whether system availability including reliability and maintainability, satisfies RAM targets and other RAM requirements.

RAM demonstration

Through the RAM demonstration process in railway projects, it is ensured that the railway system or subsystem meets the required contractual targets for RAM. This involves a systematic analysis and demonstration process that includes planning, management, and implementation procedures. The RAM demonstration aids in the specification, assessment, and potential improvement of the RAM characteristics of products and systems throughout all phases of the railway lifecycle. It is crucial for delivering a railway system that complies with applicable requirements and standards, thereby ensuring efficient and reliable operation of the railway system. Additionally, it helps identify areas for improvement and implement necessary changes to enhance the system's performance.

The RAM demonstration plan includes the following contents:

- Roles and responsibilities of the FRACAS organisation
- Definition of Service Availability, including RAM target and formula
- Product Breakdown Structure (PBS)
- Definition of service failure
- Recording items and processes
- Evaluation process for failures and causes
- Correction process
- Reporting process
- Database management and statistics

Below is a simplified example of operation and downtime record for an equipment:

Figure 57 - Records of operation & downtime

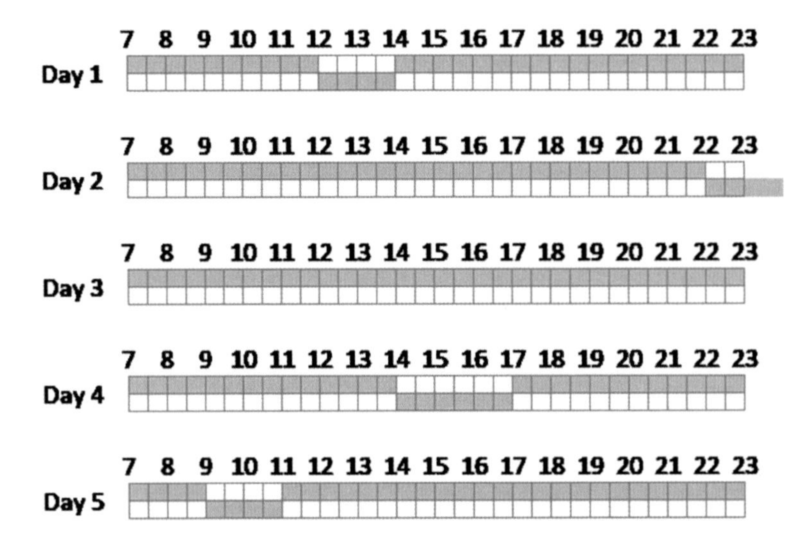

- Upper grey cells indicate that the equipment is operational, while the lower grey cells represent downtime.
- Daily operation time is from 7 a.m. to 11 p.m. (totally 16 hours).
- In Day 2, the equipment experienced downtime at 11 p.m., and the repair continued for an additional 2 hours after operations closed. However, the repair time outside of operational hours is not included in the total downtime.
- There was a total of 4 failures recorded over the 5-day period.

RAM calculation is follows:

- Total operation time for 5 days is 80 hours (= 16 hours x 5 days).
- Total downtime is 8 hours (= 2 hours + 1 hour + 3 hours + 2 hours).
- MTBF is 20 hours (= 80 hours / 4 times).
- Failure rate is 0.05 per hour (= 4 times / 80 hours).
- MTTR is 2 hours (= 8 hours / 4 times).
- Availability is 90.0% (= (80 hours – 8 hours) / 80 hours) during the specified time.

FRACAS

FRACAS, which stands for Failure Reporting, Analysis, and Corrective Action System, is a framework that provides a disciplined and iterative method for addressing reliability and maintenance issues throughout the life cycle of a railway system.
The primary purpose of FRACAS in railway projects is to promote and improve the reliability and maintainability of a system. It provides a closed-loop process for resolving RAM issues. FRACAS activities encompass capturing and recording information about failures and other reliability issues, selecting, analysing, and prioritising those failures and issues, identifying corrective actions and implementing them to prevent future occurrences of those failures, and tracking reliability and maintainability information over time for further analysis.
Implementing FRACAS can yield to several benefits, including promoting reliability growth in systems, facilitating understanding of failures and their impacts, revealing inherent or underlying reliability issues, and validating or disproving the original engineering baselines established by a Failure Modes, Effects, and Criticality Analysis (FMECA).

The failure data includes the following information:

- Identification number
- Date of failure discovery
- Asset identification based on PBS or LBS
- General problem description
- Failure description defining the specific component
- Cause description
- Remediating action description
- Indicator for service-interrupting failure (Yes/No)

The corrective action data includes the following information:

- Identification number
- Date the corrective action was identified
- Reference to the source of the corrective action
- Description of the corrective action
- Impact on RAM

The preventive maintenance (PM) data includes the following information:

- Identification number
- PM identification
- PM procedure reference
- Asset identification
- PM scheduled interval
- Personnel performing PM
- Total downtime to perform PM task
- Total labour hours to perform PM task
- Details on materials consumption
- Details on inspection records
- Description of faults or failures discovered

The corrective maintenance (CM) data includes the following information:

- Identification number
- Date and time of failure discovery
- Date and time of failure resolution
- Total labour hours required to resolve failure
- Details on materials consumption

The FRACAS report includes the following information:

- Corrective action ID
- Description of corrective action
- Implementation date and time
- Impact on RAM target
- Expected improvement
- Trend status

Much data is required to be recorded in performing a FRACAS; therefore, a database should be established first.

SAFETY MANAGEMENT

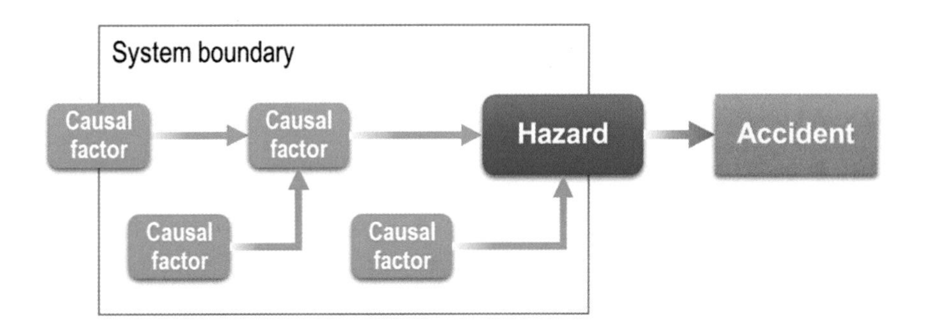

Overview

In project implementation, two types of events may occur: planned events and unexpected ones, the latter typically referred to as risks. From a Project Management standpoint, risks are unforeseen events that can impact a project positively, negatively, or both.
From a Systems Engineering perspective, risks are events or accidents that may lead to injury, loss of life, or infrastructure damage. Preventing such risks involves identifying root causes throughout the production lifecycle and working to eliminate or minimise them. These activities are encompassed by safety management, which applies engineering knowledge and skills to identify hazards, eliminate them, and reduce any associated risks that cannot be eliminated.

Types of risks

In safety and RAM management, risks are typically categorised as follows:

- Systematic failure: This type of failure results from inadequate or incorrect function leading to hazards, often stemming from design flaws. Systematic failures occur consistently or repeatedly and are typically rooted in human errors during the design phase.
- Random failure: This involves unexpected and unpredictable failures or malfunctions due to material weaknesses, manufacturing defects, or unforeseen environmental conditions.

Further risk categories may include:

- Human errors: Mistakes or unintended actions made by individuals that can lead to undesired outcomes or failures.
- Physical security failures: Risks from breaches in physical security, such as intentional attacks on railway infrastructure and systems.
- Cybersecurity failures: Risks from IT network security breaches, including hacking and other cyberattacks on railway systems.
- Intended failures: Risks associated with the SOTIF (Safety of the Intended Failure), where hazards arise from the system performing as intended but in an unexpected or hazardous manner.

Examples for each risk type are shown in Figure 58:

Figure 58 – Examples for risk types

Risk type	Category of source	Causal factor	Example
Systematic failures	System itself	Poorly designed system without function to prevent hazards or having malfunction to cause hazards	Train doors with hard and sharp edges
Random failures	System itself	Inherences failure rate of equipment	Irregular failure of a device
Human errors	Operators, maintainers, and passengers	Incorrect use by operators, maintainers, and passengers	A driver in a train cab presses a wrong button by mistake
Cyber security failures (IT security)	Attackers outside the system	An attack via IT network	Railway signalling disturbance due to hacking signalling systems
Physical security failures	Attackers outside the system	An attack on railway infrastructure and systems	Access of unauthorised person to OCC (Operating Control Centre)
Intended failures (SOTIF-related)	Surrounding environment	Technological limitation for unforeseen situations and scenarios	A driving accident involving a self-driving car unable to deal with unforeseen situations

- Systematic failures: Mitigated by applying safety management methodologies, requirement management, and V&V (Verification & Validation) techniques.
- Random failures: Managed using RAM methodologies, which are closely linked to reliability.
- Human errors: Often result from factors such as fatigue, lack of knowledge, miscommunication, or inadequate procedures.
- Physical security failures: Prevented through safety management techniques aimed at protecting against attacks.
- Cybersecurity failures: Addressed by implementing software and network management alongside cyber security techniques.
- SOTIF approach: Not covered in this book, as railway systems rarely face significant environmental changes in daily operations.

Hazards and accidents

A hazard is a condition that could lead to an accident, and an accident is an event that injures someone and/or damages something. The relationships between causal factors, hazard and accident are illustrated in Figure 59. The difference between an accident and an incident is that an accident is an unintended event resulting in damage, injury or harm, whereas an incident is an unintended event that may not lead to damage, harm or injury.

Figure 59 – Relationships between causal factors, hazards, and accidents

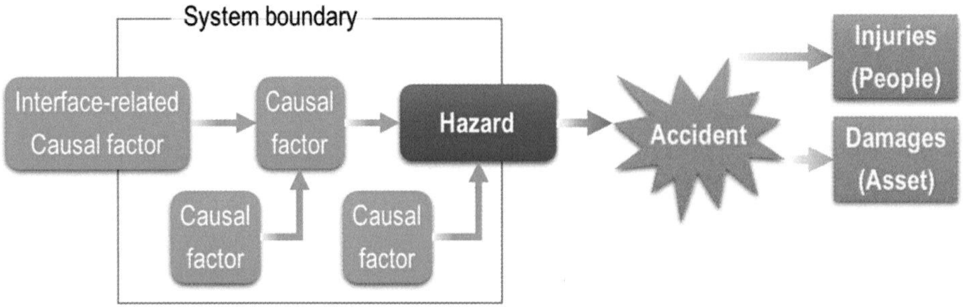

Examples of causes, hazards, accidents, and consequences are as follows:

Cause	Hazard	Accident	Consequence
Use of water	Wet floor	Slip	Injured
Power cable exposed to the outside	Aged cable	Electrocution	Injured
Use of knife	Misuse of knife	Cut	Injured

Sometimes causal factors and hazards are confused. The way to distinguish between them is to ask the question: "Is it directly associated with an accident that can injure someone and/or damage something?" If something is directly related to an accident, it is classified as a hazard; if not, it is considered a causal factor. Hazards can be prevented or mitigated, while causal factors cannot be eliminated, as they are necessary conditions or activities for operations.

Risk management process and risk evaluation

Risk management is a technique to manage hazards with the following process:

- Identify hazards and their risk factors (severity and frequency) that have the potential to cause harm. HAZOP (HAZard and OPerability study) can be conducted as part of a Quantitative Risk Analysis (QRA) for risk assessment. It is a well-documented process to analyse hazards by involving multi-disciplinary teams in the process to search for deviations from design intent.

- Analyse and evaluate the risks associated with accidents by using a risk matrix.
- Determine appropriate ways to eliminate the hazard or to control the risk that cannot be eliminated. All risks and relevant information, including mitigation measures, should be registered in the hazard log.
- Monitor risks, update the hazard log, and identify new hazards by conducting the risk management process.

Generally, risks are analysed in terms of frequency and impact (severity), with two methods commonly used to determine their criticality:

- Criticality is calculated by multiplying frequency and severity. This method is usually used in project management for managing all risks in terms of cost.
- Criticality is defined based on a risk matrix, which is normally employed in safety management.

< Safety analysis methodologies >
Safety analysis methodologies can be classified as follows:

Qualitative analysis: PHA, SHA, IHA, OSHA, DSA, and SIL analysis

- Identify safety requirements and estimate the risk in a cost-effective manner.
- Demonstrate safety verification and validation.
- Needs to be supported by quantitative analysis for high-risk concerns.

Quantitative analysis: risk matrix, FMECA

- Define risk frequency and severity.
- Evaluate risks.

Target-oriented analysis: Fault Tree Analysis, Event Tree Analysis, and Cost-Benefit Analysis

- Perform an assessment of the effect of mitigation measures and combinations.
- Demonstrate compliance with qualified risk targets.

Safety Integrity Level (SIL)

SIL is defined as rating system of risk reduction provided by a safety function, or to specify a target level of risk reduction. In rail transport, SIL is a measure of the required reliability and performance of safety-related systems to reduce the risk of failures. It indicates the level of risk reduction needed for a particular function, with SIL 1 being the lowest and SIL 4 the highest. Each SIL level corresponds to a probability range of acceptable failure rates. The higher the SIL, the more stringent the safety requirements are. Implementing the appropriate SIL ensures that safety systems are robust enough to protect against potential hazards, thereby enhancing overall system safety. SIL can be considered a part of safety requirements.

EN 50126-2 introduces recommended techniques for each SIL, as shown in Figure 60, where "H" means "Highly recommended", and "R" signifies "Recommended". For more detailed information regarding SIL, please refer to EN 50129.

Figure 60 - Recommended techniques for each SIL

Technique/Measure	SIL 1	SIL 2	SIL 3	SIL 4
Preliminary hazard analysis	H	H	H	H
Fault tree analysis	R	R	H	H
Markov diagrams	R	R	H	H
FMECA	R	R	H	H
HAZOP	R	R	H	H
Cauas-consequence diagrams	R	R	H	H
Event tree	R	R	R	R
Reliability block diagram	R	R	R	R
Zonal analysis	R	R	R	R
Interface hazard analysis	R	R	H	H
Common cause failure analysis	R	R	H	H
Historical event analysis	R	R	R	R
Operating and support hazard analysis	R	R	R	R

ISA (Independent Safety Assessment)

An Independent Safety Assessment (ISA) is an objective evaluation of the safety aspects of a system, process, or product conducted by an external and impartial party. The goal of an ISA is to ensure that the system meets regulatory safety standards and operates without posing undue risks.
Since an ISA provides an unbiased evaluation of the safety of a system, clients prefer to have an ISA to ensure system safety. Here is a breakdown of what an ISA typically involves:

- Objective evaluation: An ISA is conducted by independent experts who have no direct involvement in the design, implementation, or operation of the system being assessed.

- Review of safety requirements: The ISA team examines safety requirements and standards relevant to the system under assessment.
- Hazard identification and analysis: The team conducts a thorough analysis to identify potential hazards associated with the system.
- Risk assessment: Based on the identified hazards, the ISA team assesses the associated risks.
- Recommendations for improvement: Following the assessment, the ISA team provides recommendations for mitigating identified risks and enhancing the overall safety of the system.
- Documentation and reporting: The findings of the ISA are documented in a comprehensive report.

Because of the ISA team's activities, a safety manager in the systems engineering team should maintain close communication with the ISA team.

Safety analysis

Safety management encompasses the engineering knowledge and skills to identify hazards and subsequently eliminate or reduce the safety hazards associated with a system. In systems engineering, safety management involves identifying potential safety hazards, evaluating the associated risk of harm for each hazard, and implementing measures to mitigate these risks.
The main objectives of safety management in systems engineering are as follows:

- To minimise the risk of accidents, incidents, or harm to people, property, or the environment.
- To comply with relevant safety regulations, standards, and guidelines.
- To optimise the design by considering the safety of the system and its impact on the environment.
- To continuously monitor and review the safety performance of systems and implement measures to improve safety as necessary.

Safety management in systems engineering is a critical aspect of the systems engineering process and should be considered throughout the entire life cycle of a system, from the conceptual design phase to the decommissioning of the system. It requires the active participation and collaboration of all stakeholders, including system engineers, safety engineers, operational personnel, and regulatory bodies.

Hazard identification

Preliminary hazards list (PHL) should be developed at the beginning of the project. To create a PHL, a HAZID (HAZard IDentification) workshop can be conducted. This approach applies HAZOP for hazard identification within a railway system. HAZID can be performed across various hazard identification activities, such as PHA, SHA/SSHA, IHA, and O&SHA.
When performing a workshop for hazard identification, the identified hazards should be analysed in terms of frequency and impact, and mitigation measures should also be developed.

< HAZID >
HAZID is a systematic process used for identifying potential hazards in a facility, operation, or project. Its main purpose is to recognise and evaluate hazards early to implement measures that mitigate or eliminate risks. This involves:

- Gathering information,
- Identifying hazards through brainstorming or structured reviews,
- Analysing and classifying these hazards,
- Documenting the findings, and
- Following up with appropriate risk management actions.

The process of HAZID is as follows:

- Appoint a workshop leader and secretary
- Define objectives and scope
- Select the workshop team members
- Arrange the schedule and venue for workshop meetings
- Obtain suitable data
- Plan the sequence for the workshop
- Prepare the guidance notes
- Carry out the HAZID workshop
- Produce a HAZID report
- Track actions programmed in the report

Data for the workshop includes the following:

- Design or layout drawings
- Functional flow diagram
- Design schematic diagram
- O&M process diagrams and O&M manuals from suppliers for generic products or drafts for system level
- Equipment schedule

HAZID is a higher-level process than HAZOP or SWIFT.

< HAZOP >

HAZOP (HAZard and OPerability) is a systematic and structured technique used to identify potential hazards and operational issues. HAZOP review helps engineers ensure that all potential gaps between the hazards and risk assessment process are identified.

Key aspects of HAZOP are:

- Objective: To identify potential hazards and operability problems that could affect safety, efficiency, and performance.
- Method: Conducted through a detailed and systematic review of the process design and operation, typically involving a multidisciplinary team.
- Guidewords: Predefined guide words like "No", "More", "Less" and "Part of" are used to explore deviations from normal operations and their potential impacts (See Figure 61).
- Team: Involves experts from different fields such as engineering, operations, and safety to provide diverse perspectives and insights.
- Outcome: Results in a list of identified hazards and issues, along with recommendations for mitigating risks and improving safety and operability.

Figure 61 – Guidewords and examples

Guidewords for HAZOP	Meaning	Examples
No	Not performing	A wayside signal does not work when a train moves into the block.
More	Performing more than criteria	The red light of the signal still remains turned on for a long time, even though a train has moved to another block.
Less	Performing less than criteria	The red light of the signal turns on too slow after a train moves into the block.
As well as	Other parts are also performing	The green, yellow and red light of the signal turn on at the same time.
Part of	Only a part of device is performing	Only the yellow light of the signal work.
Other than	Performing in different ways	The green light turns on when a train enters the block.

< SWIFT >

SWIFT (Structured What-IF Technique) is a systematic and organised approach used in hazard identification and risk assessment. It involves a facilitated brainstorming session, typically led by a trained facilitator, in which participants ask structured "What if?" questions about various aspects of a process, system, or activity to identify potential hazards, risks, and failure scenarios.

Here's a process for conducting a Structured What-If Technique (SWIFT) session:

- Preparation: Assemble a multidisciplinary team with expertise relevant to the system or process being analysed. Prepare documentation and diagrams, such as process flowcharts, to provide context.
- Define the scope: Clearly define the boundaries of the analysis, including the system, activity, or process to be examined. Specify what aspects are within and outside the scope of the assessment.
- The facilitator leads the session: A trained facilitator guides the session using a set of structured "What if?" questions. The facilitator may also use guidewords, such as "failure of," "more than," or "less than," to frame the questions and ensure comprehensive coverage.
- Brainstorming: The team collectively responds to each "What if?" question, brainstorming potential hazards, their causes, and possible consequences. All identified hazards are documented in real-time.
- Documentation and reporting: The results of the SWIFT session, including identified hazards, risk assessments, and recommended actions, are compiled into a report for further review and implementation.

< Hazard log >

A hazard log is a continually updated record of identified hazards. It serves as a document or database that records potential safety hazards, their associated risk levels, and relevant information. A hazard log typically includes the following information for each hazard:

- Hazard description: A brief description of the hazard and its location.
- Hazard category: The type of hazard, such as an operational hazard or maintenance hazard.
- Hazard risk level: The severity and likelihood of harm associated with the hazard, typically determined through a risk assessment process.
- Mitigation measures: The measures implemented to reduce the risk of harm associated with the hazard, such as changes to procedures, the use of personal protective equipment, or the implementation of safety training programs.
- Responsible party: The person or team responsible for implementing the mitigation measures and monitoring the hazard.
- Status: The status of the hazard, such as open, closed, or under review.

The hazard log is used to track and manage hazards over time, ensuring that all hazards are evaluated and addressed in a consistent and systematic manner. It should be regularly reviewed and updated.

Preliminary Hazard Analysis (PHA)

PHA is the first step in the system safety process to identify and categorise hazards or potential hazards relating to the operation of a proposed system.

As introduced in the chapter on "Hazard identification", HAZID and HAZOP can be used to identify hazards. One effective methodology for HAZID and HAZOP is the fish-bone diagram, as shown in Figure 62, which helps engineers generate ideas for identifying hazards.

Figure 62 – An example of fish-bone diagram

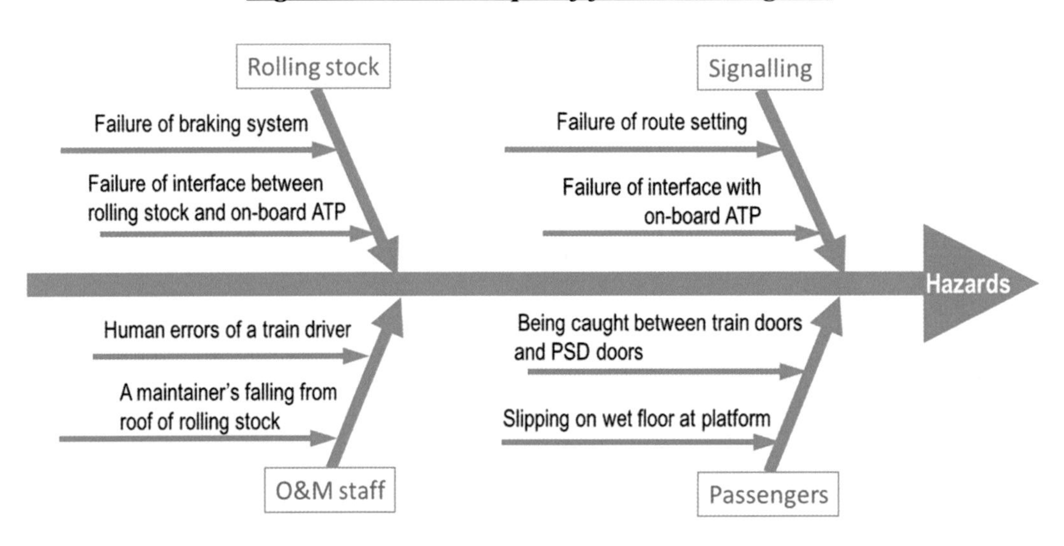

Input for conducting PHA includes the following:

- Project hazard list (if applicable)
- Preliminary hazard list
- PBS or LBS
- Graphical materials such as fish-bone diagram, system data flow, FBD (Functional Block Diagram), etc.

Activities in conducting PHA include:

- Identifying hazards, causes, and consequences
- Evaluating risks
- Developing mitigation measures

The output of PHA can be used:

- To define SSRS (System Safety Requirements Specification) and interface requirements from mitigation measures, which need to be assigned to individual systems and subsystems.
- As hazard items in SHA (System Hazard Analysis), IHA (Interface Hazard Analysis), and OSHA (Operation & Support Hazard Analysis).

The PHA worksheet should include the following data at a minimum:

- Hazard ID: The preliminary hazard identification number.
- Hazard description: A situation or circumstance with the potential of an accident that may cause injury or fatality to personnel, or damage to the system or environment.
- Operating mode: Defines the mode of operation, such as normal, degraded, emergency, and maintenance, relevant to the hazard under consideration.
- Location: The location where the hazard could occur.
- Potential cause: The failure, event, root cause that immediately leads to the hazard being realised.
- Involved sub-system: Identification of sub-systems relevant to the hazard under consideration.
- Potential accident: The type of potential accident.
- Effect/consequence: Description of the consequences of hazard.
- Severity level: The severity level before confirmation of the mitigation measures.
- Provision/safety measure: Measures that control the risk, such as design safeguards and construction procedures for risk mitigation.
- Remarks: Additional information or relevant notes.

System Hazard Analysis (SHA)

SHA or SSHA (Subsystem Hazard Analysis) is the step to conduct a detailed study of hazards at the subsystem or system levels. To perform SSHA, the preliminary hazard list (PHL) and the preliminary hazard analysis (PHA) should be prepared first.
As introduced in the chapter “Hazard identification”, HAZID and HAZOP can be used in this analysis to identify hazards. One effective methodology for HAZID and HAZOP is LBS-Guidewords matrix, as shown in Figure 63, which assists engineers in generating ideas for identifying hazards.

Figure 63 - Syste-guidewords matrix

			Guidewords for systems																Guidewords for human factors									
System	Subsystem	Component	Excessive	Gap between	Operating even while	Not operating	Failure	Collapse	Aerodynamic force	Defective	Unauthorised operating	Overspeed	Unstable	Unwarranted	Inaddequate integrity of	Pollution	No ergonomics applied	Not locked	Confused or distracted	Trapped inside	Trip, slip or fall	Exposure to	Cut	Detraining	Trapped inside	Unauthorised behaviour	Lack of information	Manual handling injury
Rolling stock	Carbody	Frame						x	x				x															
		Interior equipment													x		x						x			x		x
		Exterior equipment																										
		Doors	x	x		x									x			x		x						x	x	
	Bogie	Frame						x																				
		Bolster																										
		Wheel set																										
	Traction	Motor			x		x				x	x																
		Brake			x		x			x				x		x												

Input for conducting SHA or SSHA includes the following:

- Functional requirements
- Functional list
- PHL and output of PHA
- Experience of specialists (e.g., EMC, noise & vibration, human factor, fire, etc.)
- Project hazard list (if applicable)

Activities in conducting SHA or SSHA include:

- Identify hazards, causes, and consequences
- Evaluate risks
- Develop mitigation measures

The output of SHA or SSHA can be used:

- To define safety requirements based on mitigation measures
- As input for design changes.

SHA worksheet will contain similar data to the PHA worksheet, with unique information as follows:

- SSHA ID: The subsystem identification number.
- Original risk: The original risk before confirmation of the mitigation measures.
- Proposed mitigation measures: The measures that control the risk.
- Residual risk: The risk ranking after confirmation of the mitigation measures.
- Hazard controller: The responsible party who owns the hazard.
- Change history: Any changes to hazard information.

IHA and OSHA

< Interface Hazard Analysis (IHA) >
IHA is the technique used for addressing possible interfaces between subsystems.

Input for conducting IHA includes:

- Interface list (or interface matrix)
- Project hazard list (if applicable)
- Output of PHA

Activities in conducting IHA include:

- Identify hazards, causes, and consequences
- Evaluate risk
- Develop mitigation measures

Output of IHA can be used:

- To define safety requirements from mitigation measures
- To assign interface parties and transfer responsibilities (whole hazard or sharing mitigation measures)
- As input for design changes.

< Operating & Support Hazard Analysis (OSHA) >
OSHA is the technique used to identify all hazards in the operation of a system that are inherently dangerous to personnel or where human errors could pose hazards to equipment or people.

Input for conducting OSHA includes the following:

- Operational scenarios
- Suppliers' O&M (Operation and Maintenance) manual
- System OSHA list
- Output of SHA/SSHA and IHA

Activities involved in conducting OSHA are:

- Identify hazards, causes, and consequences
- Evaluate risks
- Develop mitigation measures

The output of OSHA can be used:

- To develop an emergency plan
- To define training requirements
- To provide input for system O&M manual
- To inform design changes

Safety management techniques

Safety-related principles and requirements

< Safety-related principles >

There are many principles relating to safety that can be applied to system design as follows:

- Fail-safe principle: Ensures that a system is designed to automatically revert to a safe state in the event of a failure or error. In other words, if something goes wrong, the system will default to a state that prevents further harm or damage.
- Checked-redundancy principle: Involves having multiple components perform the same function, then comparing their outputs to detect errors.
- Collective protection principle: Prioritises measures that protect groups of people over individual protective measures.
- Safety margin principle: Refers to building in extra capacity or tolerance beyond the expected load or stress to account for unexpected conditions or failures.
- Defensive design principle: Involves anticipating potential failures or misuse and designing the system to handle these situations safely.

< Safety requirements >

- Safety target is the numeric target relating to safety, such as the annual number of injured or deceased passengers, for instance.
- Safety requirements are criteria that a system must satisfy to be deemed acceptably safe, given applicable inputs and conditions. Satisfying the safety requirements involves demonstrating through evidence that safety risks associated with the system have been managed throughout the system lifecycle and reduced to an acceptable level.
- Safety Integrity requirements are assigned to safety functions.

Bow-tie method

The bow-tie method in safety management is a risk assessment and management tool that visualises relationships between potential causes of a hazard, the hazard itself, and the consequences of that hazard. The name “bowtie” comes from the shape of the diagram, which resembles a bowtie, as shown in Figure 64.

Figure 64 – Bowtie diagram

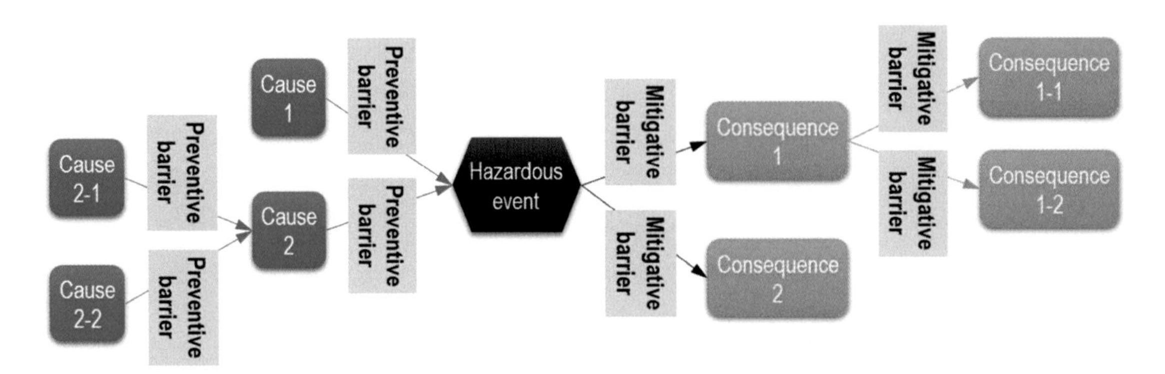

There are some key components:

- Hazardous event: The central event or hazardous situation that could potentially lead to undesirable consequences.
- Causes (or threats): Events or conditions that could lead to the hazardous event.
- Sub-causes (or sub-threats): Events or conditions that could lead to the cause.
- Preventive barriers: Measures placed between the causes and the hazardous event to prevent the occurrence of the hazard or the cause.
- Consequences: Potential outcomes or impacts if the hazard materialises.
- Sub-consequences: Secondary outcomes or impacts if the consequence materialises.
- Mitigative barriers: Measures placed between the hazard and the consequences to mitigate the impact if the hazard or the consequence occurs.

ALARP (As Low AS Reasonably Practicable)

ALARP is a principle in the management of safety-critical and safety-involved systems. The principle states that the residual risk should be reduced as low as reasonably practicable.
Through FMECA (Failure Modes, Effects, and Criticality Analysis), each subsystem is analysed and assigned a risk ranking based on a predefined risk matrix as shown in Figure 65. The safety level of components is evaluated based on the risk matrix. If a component is analysed and found to have an "Occasional" degree in frequency and "Critical" in severity, it is assigned an "R3" risk ranking.

Figure 65 – An example of risk matrix

		Severity			
		Insignificant	Marginal	Critical	Catastrophic
Frequency	Frequent	R3	R4	R4	R4
	Probable	R2	R3	R4	R4
	Occasional	R2	R2	R3	R4
	Remote	R1	R2	R2	R3
	Improbable	R1	R1	R2	R2
	Incredible	R1	R1	R1	R2

From the standpoint of the overall system level, however, the component is not desirable. Engineers should try to reduce the risk level (risk ranking) of the component to "R1." It is essential to keep in mind that reducing risks always requires cost or effort. The ALARP principle provides guidance on how much effort engineers should invest in reducing risks.

In the above case, any component ranked at "R3" or "R2" will be the ones whose risks should be reduced further. Methodologies to reduce risks include decreasing failure frequency and/or decreasing severity.

ALARP will be applied except when:

- No further mitigation measures are implementable.
- The cost for mitigation measures is significantly greater than the benefits from derived from their implementation.

SFAIRP (So Far As Is Reasonably Practicable)

While ALARP is performed based on practicality and cost-effectiveness, SFAIRP involves evaluating what measures are reasonably practicable.

Methodology for implementing SFAIRP is:

- Hazard identification: Conduct a thorough analysis to identify all potential hazards of the system.
- Risk assessment: Evaluate the risks associated with the identified hazards and determine the likelihood and severity of each risk.

- Control measures: Identify all possible measures that could mitigate or eliminate the risks, considering engineering controls, administrative controls, and personal protective equipment (PPE).
- Reasonableness evaluation: Assess the feasibility and effectiveness of each control measure, considering factors such as cost, practicality, and proportionality of the measure related to the risk.
- Controls implementation: Implement the identified control measures that are deemed reasonably practicable.
- Documentation: Maintain detailed records of the hazard identification, risk assessment, control measures, and the justification for the decision made.

Safety Management Plan

Scope of work

Defining the scope of work clearly in a safety management plan is crucial because it establishes the specific tasks, boundaries, and responsibilities for all involved parties. This helps to identify potential hazards, allocate resources effectively, and ensure that all safety measures are tailored to the exact nature of the work being performed. A well-defined scope minimises misunderstandings, reduces the risk of accidents, and ensures compliance with safety regulations.
Another important aspect of defining the scope of work is that a clarified boundary helps delineate responsibilities between sub-contractors. Causal factors can reside inside, outside, or at the system boundary, and any vagueness in this boundary can lead to disputes between sub-contractors.

< Presumptions and prerequisites >
Defining presumptions and prerequisites in a safety management plan ensures that all parties understand the baseline expectations, such as required training, certifications, or environmental conditions, before work begins. When developing a plan, an author needs to have some inputs to reflect on the plans. However, if the necessary input data is not obtained at the time of writing the plan, all of these should be described in the "Presumptions and Prerequisites" section of the plan. Any concerns that the author has regarding the lack of information for developing the plan, should also be detailed in the section.

Safety management activities and methodologies

Input for developing safety assurance activities and methodologies includes the following:

- Bid proposal
- Contractual document
- Project-wide RAMS plan (if applicable)
- Project plan & milestone programme
- Standards and guidance

Safety management activities and methodologies represent the safety analysis activities introduced in the chapter "ISA (Independent Safety Assessment)

An Independent Safety Assessment (ISA) is an objective evaluation of the safety aspects of a system, process, or product conducted by an external and impartial party. The goal of an ISA is to ensure that the system meets regulatory safety standards and operates without posing undue risks.

Since an ISA provides an unbiased evaluation of the safety of a system, clients prefer to have an ISA to ensure system safety. Here is a breakdown of what an ISA typically involves:

- Objective evaluation: An ISA is conducted by independent experts who have no direct involvement in the design, implementation, or operation of the system being assessed.
- Review of safety requirements: The ISA team examines safety requirements and standards relevant to the system under assessment.
- Hazard identification and analysis: The team conducts a thorough analysis to identify potential hazards associated with the system.
- Risk assessment: Based on the identified hazards, the ISA team assesses the associated risks.
- Recommendations for improvement: Following the assessment, the ISA team provides recommendations for mitigating identified risks and enhancing the overall safety of the system.
- Documentation and reporting: The findings of the ISA are documented in a comprehensive report.

Because of the ISA team's activities, a safety manager in the systems engineering team should maintain close communication with the ISA team.

Safety analysis." The activities are as follows:

- Translate into demonstrable targets (e.g., tolerable risk, ALARP)
- Review standards
- Conduct analysis
- Define the mechanism for transferring of safety responsibilities

Safety management process

Safety management phase can be divided as follows:

- Hazard identification
- Hazard database (hazard log) management
- Hazard analysis

- Application
- Verification and validation

< Hazard identification >

Hazards are identified through PHA, HAZID, and HAZOP at the beginning of a project, and all the results are recorded in the Hazard Log (hazard database). These identified hazards are used to develop Safety Requirements Specification.

< Hazard log management >

Hazard log is a database for safety management, containing all hazards identified through hazard identification activities and hazard analysis activities such as SHA, SSHA, IHA, and OSHA. It also includes safety issues raised from human factors management, interface management, and EMC management. Therefore, the hazard log should be updated throughout the project lifecycle. The hazard log will be used as a source worksheet for safety verification and validation.

< Hazard analysis >

Based on the Safety Requirements Specification (SRS), SHA is conducted. SSHA, IHA, and OSHA will be performed based on the SRS and SHA. Hazard analysis should also include quantitative analysis, such as FMECA.

< Application >

Safety requirements should be applied to system designing a system and actualisation.

< Verification and validation >

The hazard log, containing safety requirements, will be used as a requirements matrix with which an assessor verifies or validates the design and deliverables.

Safety Verification and Validation

Safety Verification and Validation (V&V) activities are collectively referred to as safety justification.

Safety audit & review and V&V

< Safety audit and review >

Safety manager should carry out periodic safety reviews and audits to monitor suppliers' safety management activities and compliance with requirements for the following purposes:

- Safety audits focus on the safety management process being used and check that it is adequate and followed. They also demonstrate that, so far as reasonably practicable, passengers, staff and others affected by the railway are not exposed to unacceptable risks to their health and safety from the operations of a line.
- Safety reviews are a type of pre-inspection work that checks overall safety management before performing safety V&V (Verification and Validation) to prevent excessive non-compliances from occurring during the V&V activities.

However, safety audits and reviews are not clearly distinguished in practice. Both can be conducted to guide suppliers in preparing for safety V&V. There is another perspective that considers safety audits and reviews as activities related to V&V.

< Safety V&V >

Safety verification involves determining through safety analysis and appropriate testing that each phase of the lifecycle satisfies the specific safety requirements identified in the previous phase.

Safety verification and validation are essential because they ensure that safety measures and systems meet all required standards and function as intended. Verification checks whether the design and implementation align with safety requirements, while validation confirms that the final product or system performs safely in real-world conditions. These processes help identify and mitigate potential safety risks, ensuring the protection of people, property, and the environment.

All safety requirements, as well as other requirements relating to EMC, human factors, and similar areas, should be managed through a requirements management process.

Safety case

A safety case is a comprehensive written demonstration of evidence and due diligence that shows a system can be operated safely and effectively control hazards. It presents a structured safety argument developed incrementally based on sound logical reasoning for acceptance. In rail transport, a safety case report specifically illustrates how a railway system, project, or component meets all safety requirements and adequately controls identified hazards. This report typically includes risk assessments, safety analyses, and compliance with relevant standards and regulations, making it crucial for ensuring stakeholder confidence, regulatory approval, and the safe operation of the railway system.

Types of safety cases for each development level are as follows:

- Safety case for generic product (GP): Provides evidence that a generic product is safe in all applications (e.g., a points machine)
- Safety case for generic application (GA): Provides evidence that a generic product is safe in a specific class of applications (e.g., the points machine used at a particular type of junction)
- Safety case for specific application (SA): Relevant to one specific application (e.g., the points machine used in a particular signalling scheme)

< Safety Case Report (the final safety report) >
A safety case report should consist of the following contents:

- System description: Detailed information about the system, including its purpose, design, and operational context.
- Hazard identification: Identification of potential hazards associated with the system or operation.
- Risk assessment: Evaluation of the risks posed by the identified hazards, including their likelihood and impacts.
- Safety measures: Description of the controls, safeguards, and procedures to mitigate identified risks.
- Evidence: Collection of data, analysis, and documentation demonstrating the effectiveness of the safety measures and compliance with safety standards.
- Review and approval: Process for reviewing the safety case, including internal and external audits, and obtaining approval from regulatory bodies.

HUMAN FACTORS

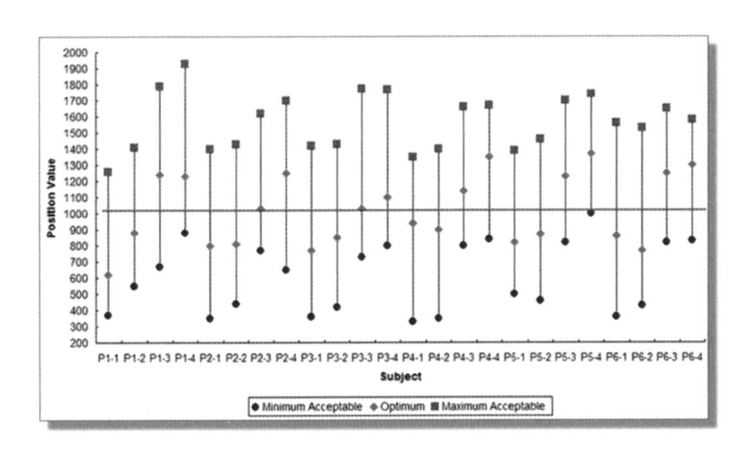

Overview of Human Factors

Human Factors (or ergonomics) is the application of psychological and physiological principles to the engineering and design of products, process, and systems to ensure safety, efficiency, and convenience in relation to human behaviour.
Human Factors aims to design products, systems, and environments that are user-friendly, safe, and efficient by considering the physical and cognitive characteristics of the users. This field of study contributes to creating products and systems that are comfortable, intuitive, and easy to use, ultimately enhancing overall human performance, satisfaction, and well-being. The goals of human factors in railway projects are as follows:

- Reducing human errors by operators, such as drivers and signallers, which can lead to accidents.
- Enhancing productivity by reducing fatigue among maintainers or users.
- Improving comfort for passengers, operators, and maintainers in interaction between humans and machines.

Cognition errors

The knowledge body of Human Factors can be classified into psychological (cognition), physical, organisational, and environmental factors. The process of human cognition based on self-judgment involves sensing stimuli and recognising what is sensed, as shown in Figure 66.

Figure 66 - Sensation and cognition process

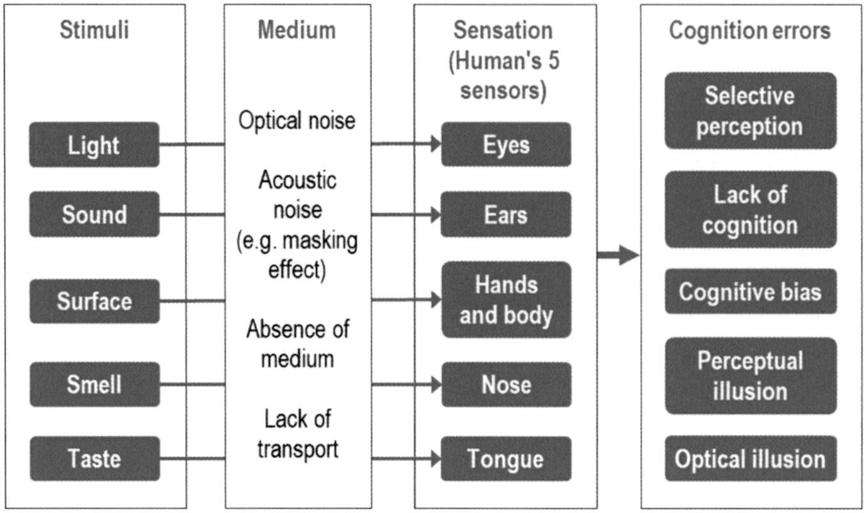

Stimuli: Light, Sound, Surface, Smell, Taste

Medium: Optical noise; Acoustic noise (e.g. masking effect); Absence of medium; Lack of transport

Sensation (Human's 5 sensors): Eyes, Ears, Hands and body, Nose, Tongue

Cognition errors: Selective perception, Lack of cognition, Cognitive bias, Perceptual illusion, Optical illusion

While recognising sensory input, human cognition can result in outcomes, as follows:

- Normal perception: Recognising without any errors.
- Selective perception: A cognitive process where individuals focus on certain stimuli while ignoring others, based on their expectations, beliefs, or experiences.
- Lack of recognition: A failure or lack of recognition due to low clarity of stimuli, interference from the media, or limitations of human sensory capabilities.
- Cognitive bias: A systematic error in thinking that occurs when people process and interpret information in their environment.
- Perceptual illusion: A distortion or misinterpretation of a sensory experience, where the perception of an object differs from the objective reality.
- Optical illusion: A visual phenomenon where the brain interprets an image differently from its actual form.

< Selective perception >

An example of selective perception is when a doctor, influenced by a preconceived notion about a particular symptom or illness, might overlook other important symptoms or possibilities. Selective perception significantly affects our everyday decision-making and cognitive process.

On the other hand, human factors engineers in railway projects should focus on the public's selective perception, rather than the individual's, as an individual's selective perception varies greatly due to personal experience.

< Lack of cognition >
Low clarity of stimuli and/or the limitations of human sensors result in a lack of recognition or failure to recognise. Significant examples of this lack of cognition include:

- Vision: Difficulty observing an object due to its small size or difficulty finding an object due to its low contrast of colour with its surroundings.
- Audio: The masking effect (the inability to detect small sounds when loud sounds occur simultaneously), and the orchestra effect (when multiple sounds are heard at the same time, making it difficult to recognise the sounds of individual instruments).

< Cognitive bias >
Cognitive biases are often the result from the brain's attempt to simplify information processing. Some types of cognitive biases include:

- Anchoring bias: The tendency to rely on heavily on the first piece of information encountered. For instance, if you learn the price of a product in a market, you may judge the price of any similar product as high or low in comparison to the initial price.
- Halo effect: Your overall impression of a person influences your perception of their characteristics.

< Perceptual illusion or optical illusion >
Although an illusion is like a lack of cognition, the difference lies in that a lock of cognition refers to an unrecognised state, while an illusion leads to a misunderstanding. One of the most notable examples of illusion is optical illusion, for which many examples can be found online.

Physical constraints

Physical constraints are related to ergonomics, anthropometry, disability, and physiology. Due to their overlap, distinguishing them clearly can be challenging.

< Ergonomics >

From the perspective of physical constraints in human factors, ergonomics involves designing tools, systems, and environments that align with the physical limitations and capabilities of the human body. This means ensuring that objects are designed to accommodate the size, strength, and range of motion of users, thereby reducing fatigue and preventing injuries. For example, a workstation might be designed with adjustable height and proper monitor positioning to prevent repetitive strain and enhance comfort and efficiency.

< Anthropometry >

Anthropometry involves the measurement and study of human body dimensions to ensure that products, workspaces, and systems are designed to fit the physical characteristics of the intended user population. By considering variations in height, reach, and strength, designers can create environments that accommodate a wide range of users, minimising discomfort and reducing the risk of injury. For instance, a chair designed using anthropometric data would feature adjustable elements to suit different body sizes and shapes, thereby enhancing usability and comfort.

< Disability >

From the perspective of physical constraints in human factors, disability refers to limitations in physical, sensory, or cognitive abilities that can affect how individuals interact with systems, products, or environments. Designing with these constraints involves creating accessible and inclusive solutions that accommodate diverse abilities. For instance, incorporating features such as ramps, adjustable interfaces, or voice-activated controls can help overcome barriers and ensure that individuals with various disabilities can use a product safely and effectively. Braille blocks serve as a good example of tools designed to overcome barriers.

< Physiology >
Physiology refers to the biological functions and capabilities of the human body, which can impact how individuals perform tasks and interact with systems. Design considerations must account for factors such as muscle strength, endurance, fatigue, and sensory limitations to prevent overexertion and ensure safety and comfort. For example, in a physically demanding task, tools and workspaces should be designed to minimise strain and allow for natural movements, thereby reducing the risk of injury and improving overall performance and efficiency.

Organisational constraints

Organisational constraints arise from laws, social customs, company regulations and processes, and the working culture within an organisation.

< Laws >
From the perspective of organisational constraints in human factors, laws refer to regulations and legal requirements that organizations must adhere to ensure the safety, health, and well-being of employees and users. These laws can dictate standards for workplace ergonomics, safety procedures, and product design, influencing how systems and processes are developed and implemented. Compliance with these regulations not only helps prevent accidents and injuries but also protects the organization from legal liabilities and enhances overall organizational effectiveness.

< Social customs >
Social customs refer to the cultural norms, traditions, and behaviours that influence how individuals and groups interact within a workplace or system. Organisations must consider these customs when designing workflows, communication systems, and policies to ensure alignment with the values and expectations of the workforce or user base. Ignoring social customs can lead to misunderstandings, reduced morale, or resistance to changes, while respecting them fosters a more harmonious, inclusive, and effective organisational environment.

< Company regulations & processes >
Company regulations and processes refer to the internal rules, procedures, and standards that guide how employees should perform their tasks and interact within the organization. These constraints can impact the design of systems, workflows, and communication channels, as they dictate acceptable practices and ensure

consistency and safety in operations. While these regulations help maintain order and compliance, they can also limit flexibility and innovation if they are overly rigid or not aligned with employees' needs and capabilities.

< Working culture in a company >

The working culture in a company refers to the shared values, attitudes, and practices that shape how employees interact, communicate, and perform their tasks. This culture can significantly influence productivity, teamwork, and job satisfaction. It can act as a constraint by establishing norms and expectations that may either support or hinder the implementation of new processes, technologies, or changes. For example, a company with a rigid hierarchical culture might struggle with open communication and innovation, whereas a collaborative culture may facilitate adaptability and employee engagement.

Environmental constraints

In a workplace, several factors must be considered for temperature, noise & vibration, pollution, and germ. These are typically well controlled in a clean office environment; however, they can be challenging to manage on a construction site or in a workshop.

< Temperature >

Temperature in a construction site or workshop can significantly affect workers' safety, comfort, and performance. Extreme temperatures, whether hot or cold, can lead to physical stress, reduce concentration, and increase the risk of accidents and errors. Effective management through proper clothing, ventilation, hydration, and rest breaks is essential to maintain worker well-being and productivity in such environments.

< Noise & vibration >

Excessive noise can lead to hearing loss, hinder communication, and increase stress levels, while vibrations from heavy machinery can cause discomfort and long-term musculoskeletal issues. Managing these factors through soundproofing, the use of personal protective equipment, and equipment maintenance is crucial for ensuring a safe and productive work environment.

< Pollution >

Pollution in a construction site or workshop refers to the presence of harmful substances, such as dust, fumes, and chemical contaminants, which can negatively affect workers' health and safety. Exposure to pollutants can lead to respiratory issues, skin irritations, and other health problems, thereby impacting overall productivity and well-being. Effective pollution management strategies, including proper ventilation, dust control measures, and the use of protective equipment, are essential to mitigate these risks and create a healthier work environment.

< Germs >

Germs can lead to illnesses among workers. Environments that are often exposed to dirt, debris, and various materials, can harbour bacteria and viruses that pose health risks. Poor sanitation, inadequate hygiene practices, and close working conditions can exacerbate the spread of germs. Implementing effective cleaning protocols, promoting good personal hygiene, and providing access to sanitation facilities are crucial measures to minimise the risk of infections and ensure worker safety and health.

Processes of Human Factors Integration

When developing a new system, the general processes of human factors (HF) integration and design development are as follows:

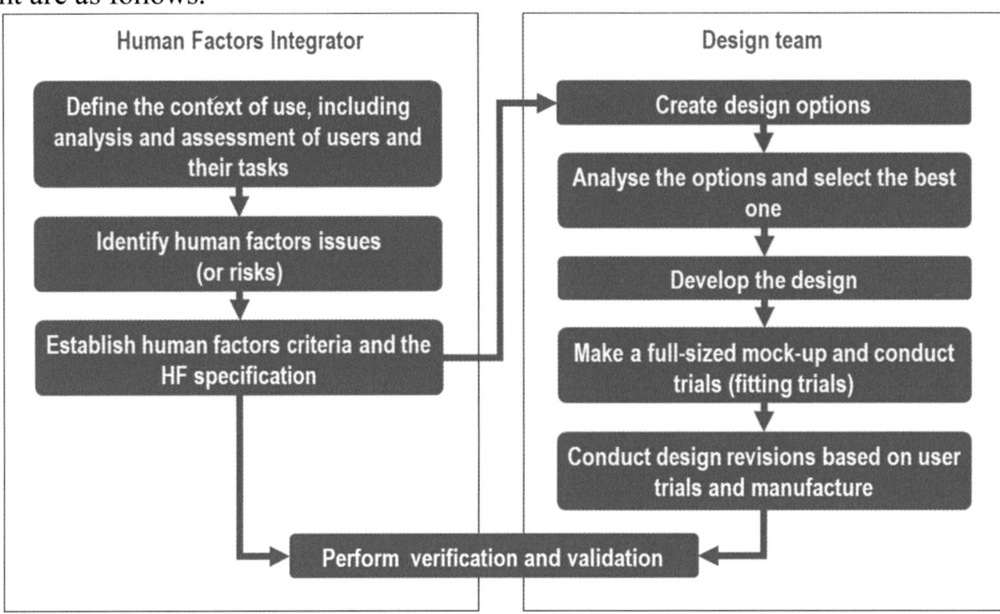

Context of use

Context of use (COU) describes the general duties, work environment, body size database (mobility and disability), training skills, qualifications, and the background of the envisaged users. The users of railway can be classified as follows:

- Primary users, who use the product or system the most (operators)
- Secondary users, who have less contact but nevertheless maintain important levels of interaction with the product or system (passengers)
- Tertiary users, who make less or infrequent use of the product (maintainers)
- Users with disabilities, including individuals who are disabled, pregnant, elderly, etc.

In railway projects, target systems include all control systems and monitor systems that require a human-machine interface. Areas may include the following:

- OCC (Operation Control Centre) and BCC (Back-up Control Centre)
- SCR (Signalling Control Centre)
- Stations, including the administration office, ticket office (passenger information office), and platform
- Substation (power supply)
- Rolling stock

< Matrix for context of use >

To analyse the context of use (COU), it is crucial to develop the matrix for COU as shown in Figure 67. It consists of user types and PBS (Product Breakdown Structure). If PBS and users are related, mark the check mark and analyse the checked items to create the COU.

Figure 67 - An example of COU matrix

Level 1	Level 2	Level 3	Operators	Maintainers	Passengers	Users with disabilitites
Station	**Admin office**	---	v	v		
		---	v	v		
	Ticket office	**Ticket desk**	v		v	
		---	v			
	Platform	**Screen doors**	v	v	v	v
		Stairs			v	
		Elevators			v	v
		Escaltors		v	v	

Identification of Human Factors (HF) criteria and issues

Human factors analysis and assessment involve evaluating how people interact with systems, products, or environments to identify potential issues that could affect performance, safety, and user experience. This process includes studying users' behaviours, cognitive processes, and physical interactions to uncover insights into usability and ergonomics. By using various methods, such as observations, surveys, and simulations, analysts can assess the compatibility of designs with human capabilities and limitations. The goal is to inform design improvements and enhance overall effectiveness, safety, and satisfaction for users. In this process, the analysis of users' tasks and risks should focus on the following two aspects:

[A] Repetitive or frequent behaviours of users.
[B] Activities where mistakes can occur.

When it comes to [A], target systems or facilities can include:

- Human-machine interface equipment – controlling and monitoring systems.
- Frequently used items – chairs, desk, etc.

Regarding [B], workflow or process should be studied. Here is a good example of mistakes that anyone can make when withdrawing money from cash machine (ATM, Automated Teller Machine):

1. Choose type of task.
2. Insert a debit card into the machine.
3. Input passwords.
4. Decide the amount of money to be withdrawn.
5. Machine processing.
6. The debit card is ejected from the ATM.
7. Take the debit card out.
8. Money is dispensed from ATM.
9. Take money.

If step 6 and step 8 are reversed, many people may forget to take their debit cards and take only their money. This easily occur
s because people tend to focus solely on what they want – withdrawing money.

< Human Factors specification >
Based on the identified criteria and issues, the human factor specifications will be developed. However, when developing human factors specification, there are two precautions to keep in mind:

- Cater only to the end users and their needs. [Fit for purpose]
- Do not focus on a loud voice. [If you concentrate on the opinions of high-ranking individuals, such as your team leader, rather than on the criteria, you may make errors in developing the specification.]

To enhance safety, the following countermeasures can be applied:

- Eliminate hazards at the source [e.g., rounded edges on the corners of toys]
- Reduce hazards at the source [e.g., proper arrangement of knives]

- Separate individuals from hazards [e.g., keeping medicine out of reach of babies]
- Contain hazards by enclosures [e.g., using covers for openings in the floor]
- Reduce employees' exposure [e.g., limiting total working hours per week]
- Wear PPE (Personal Protective Equipment) [e.g., masks]

Anything that will be interfaced with humans should be designed based on statistical anthropometry data.

Full-sized mock-up and trials (fitting trial)

When developing a new type of product or system, it is necessary to conduct a fitting trial. Figure 68 is an example of a statistical chart of the result of a fitting trial for pressing buttons on a train ticket vending machine.

Figure 68 - Fitting trial for vending machine

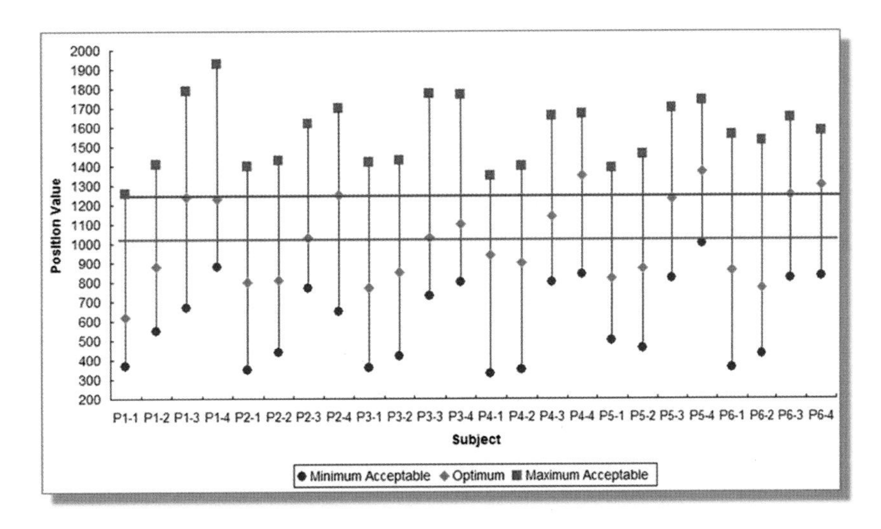

Full-sized mock-ups and trials are essential in human factors because they provide a realistic representation of products or systems, allowing users to interact with them in a controlled environment. This hands-on experience helps identify potential usability issues, ergonomic challenges, and design flaws that might not be apparent in theoretical models or smaller prototypes. By conducting trials with potential users, designers can gather valuable feedback, refine their designs, and ensure that the final product meets the needs and expectations of the intended audience, ultimately enhancing safety and user satisfaction.

< 3D modelling >

Recently 3D modelling techniques are used instead of building a full-sized mock-up to check human-machine interface as shown in Figure 69.

Figure 69 - Male reach to passenger assistance office tray

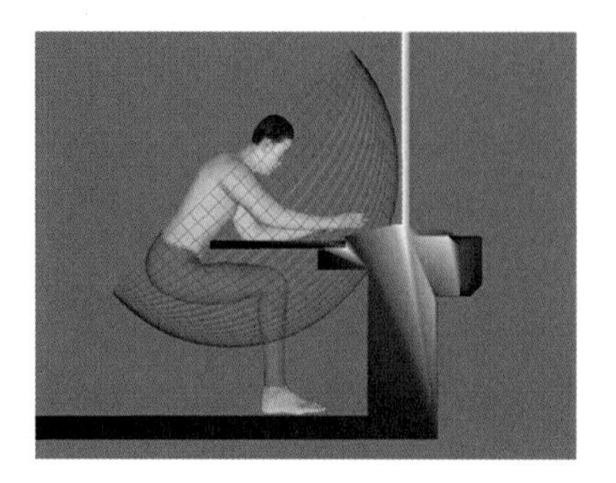

3D modelling offers several advantages over full-sized mock-ups and trials. It allows for rapid iteration and visualization of designs without the time and resource constraints of physical prototypes. Designers can easily modify and experiment with different configurations, materials, and features. Additionally, 3D models can be analysed for ergonomics and usability through simulations, enabling early identification of potential issues. This approach is often more cost-effective and can facilitate collaboration among teams, as models can be shared and reviewed digitally, enhancing the design process.

Human Factors Integration Plan

The Human Factors Integration Plan outlines the Human Factors (HF) work plan, activities, requirements, management processes, and how HF will be integrated into the system assurance and shown to be effective in meeting the applicable requirements. For an operational system to deliver the expected levels of benefits, it is essential that the human interactions with the system and its elements are well designed through the application of established HF principles and knowledge. The application of Human Factors Integration (HFI) provides a framework for achieving well-developed human-system interactions in system design. As a result, this human-system interaction can contribute to optimised system performance, and identification and mitigation of risks.

Scope of work

From the perspective of Human Factors, the HF work should cover the areas and systems where close interaction between humans and systems is required for safety, operability, and maintainability of a railway project. Figure 70 illustrates the categories of HF target systems and facilities. The details of the systems and facilities in railway projects are as follows:

- [1] Controlling systems: rolling stock, signalling, communication, power supply, OCC (Operation Control Centre), station, sub-station
- [2] Monitoring systems: large display boards in OCC, monitors in O&M (Operation and Maintenance) offices, on-board system in rolling stock, CCTV surveillance system, SCADA
- [3] Information: warning signs, alarm, public information system at platform, displays in rolling stock for passengers, displays in the driver's cab of rolling stock
- [4] Frequently used things: chairs and desks in offices, station ticket office, gate lines, layout of OCC and O&M offices
- [5] Operating things: PSD (Platform Screen Door)
- [6] Items for the users with disabilities: voice guidance
- [7] Items used by individuals with disabilities, the elderly, and the pregnant women: braille blocks, braille sign, elevators, feeding rooms

Figure 70 - HF target

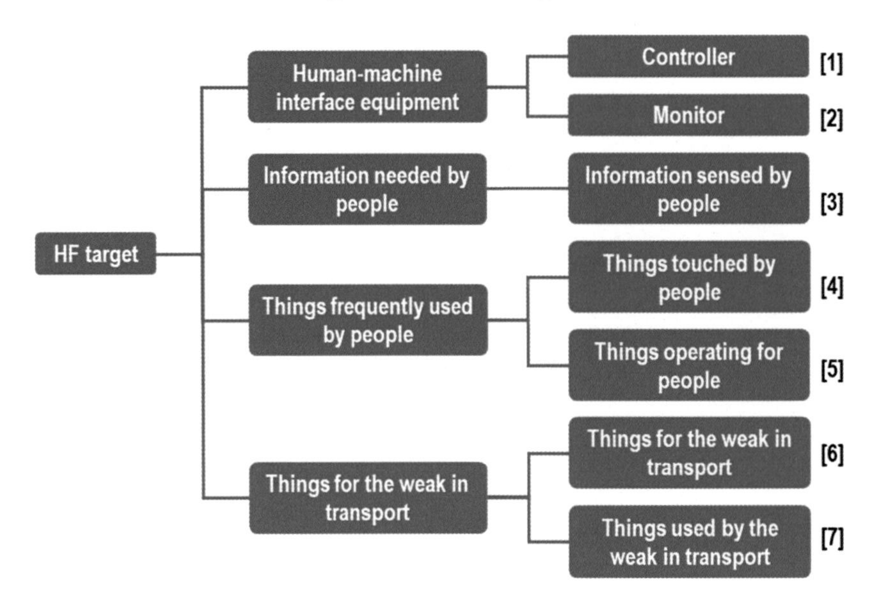

< Reference >

The following standards are worth to review when developing a Human Factors Integration Plan:

- ISO 6385: Ergonomics principles in the design of work systems
- ISO 9241: Ergonomics of human-system interaction
- ISO 10075: Ergonomic principles related to mental workload
- ISO 11064: Ergonomic design of control centre
- ISO 11226: Ergonomics – Evaluation of static working postures
- ISO 15006: Ergonomic aspects of transport information and control systems
- ISO 15534: Ergonomic design for the safety of machinery
- ISO/TR 16982: Ergonomic of human-system interaction – Usability methods supporting human-centred design

Anthropometry of each country's population should also be used as a main reference.

Human Factors engineers' roles and responsibilities

A human factors engineer is responsible for actively ensuring that human factors, usability, and human-related safety issues are fully considered in design. In practice, this involves extensive liaison within and outside the project team. Specific roles and responsibilities are as follows:

- Provide guidance and direction on HF requirements to stakeholders.
- Manage the integration of all HF issues associated with the scope of the HFIP (Human Factors Integration Plan).
- Ensure all project staff are aware of the project's HF Issues Register (HFIR) and how it will be integrated into the project.
- Ensure HF requirements are properly verified and validated.
- Coordinate the HFIP with other disciplines.
- Establish and conduct communications with the Employer's representative HF specialist (should one be assigned to the project).
- Enter and manage HF risks and risk mitigation elements in the project risk management register.

End user representatives should be involved during the design phase to ensure that their capabilities, skills, and requirements are addressed.

User identification and profiles

Identifying users and analysing their behaviours are the key processes for Human Factors task.
The main points from the user profiles in railway projects are:

- No specific personal protective equipment (PPE) should be assumed.
- Passengers may have mobility impairments, which may involve the use of a wheelchair or walking aid.
- Passengers may have colour vision deficiencies, but the railway operation staff will be able to accurately distinguish all colours.
- Passengers should be assumed to have a poor perception of the risks associated with the environment and should not be assumed to be familiar with any system.
- All railway operation staff members should be fully mobile.
- All railway operation staff will receive training associated with the systems within the OCC and stations and be familiar with the operation.

Human Factors Integration requirements

HFI aims to ensure optimisation of human-system interactions to provide effective system performance. The following are generic categories of HF requirements which are applicable to railway projects:

- Design requirements
- Anthropometric data
- Controls and displays
- Information content
- Alarms and alerts
- Control room
- Seating
- Glare and reflections
- Customers and the public

Design requirements

Design with HF requires relevant evidence and justification explaining that human capabilities, limitations, and safety are considered in various environments. The followings are the starting point of human factors activities which should be considered for designing systems and equipment:

- The context of use and the system operation.
- Human variability, to ensure the system can cater for the specific range of end users, including the range of disability.
- Human capabilities and limitations regarding physical and cognitive attributes.

For the achievement of HF-considered system design and equipment, they should be designed to be error-tolerant and to provide a means of error identification and recovery when an error occurs. They should be designed with the concept that the easiest way of doing something is the best way. They should not have distractions that impair the primary safety task, and systems designs should minimise violation. In addition, the design should consider the overall environment for all end users while adhering to good ergonomic principles.

< Anthropometric characteristics >

The physical features of human beings across the globe vary in size, strength, hearing and visual ability. Additionally, gender differences add to the scope of human variability. HF uses anthropometric data to determine suitable values for proportions, size, weight, mobility, optimised workspace layout, and the working capacity of users within the design. It is recommended that the range of anthropometric data should be from the 5th percentile female or male to the 95th percentile female or male adult for railway projects. The suitability of the data adopted for the subcontractor's design and equipment should be justified by the designer and provider with the assistance of an ergonomist in the industry. Otherwise, they should prove that their designs and equipment/facilities have a proven record of being widely adopted and employed in the industry regarding human factors concepts.

Controls and displays

- Controls and displays are associated with safety, layout, and security in the HF concept. They can also be regarded as those provided on or through a computer or other graphic user interface.
- The layout and position of controls and displays should be designed based on the principles of frequency, sequence of use, importance, and functional grouping. Primary controls should be within reach of the normal operating position, and primary displays should be visible from that same position.
- All labelling of controls and displays should be visible and legible from the expected viewing position and distances for the anticipated range of end users.
- The layout design and control selection should minimise the unintentional operation of a control.
- No control should be allowed to be placed in a mid-state position, as this does not correspond to a designed state.
- In any case where a human needs to reach a control, the design should ensure safe methods for end users to assess them, regardless of how infrequently the controls are used.
- Only the most relevant information should always be displayed. Any critical information related to safety should be readily accessible whenever necessary.
- Displays should provide information or alarms for a sufficient duration to ensure that operator or end users can detect them at any time.
- The number of colours used within HMIs should be kept to a minimum to ensure clarity.
- Specific colours should be designated for specific purposes of information delivery; for instance, red should indicate upset conditions only.
- Combination of colours that become unrecognizable should be prohibited, such as green/yellow, and white/cyan.

- The contrast on displays between the background and the foreground objects should be sufficient for recognition. It is recommended that a grey background be used to reduce glare, providing optimal conditions for discrimination of different colours in a properly lit control room.

Information content

Information content displayed should be visible, legible, and intelligible from the expected viewing distances and positions. "Intelligible" means that information content displayed should be understandable. The type of displayed information, the level of information required, the expected viewing distance, and the anticipated viewers determine the size, language, and type of the display. All information provided should be unambiguous enough to assist end users in effectively working with or using the system in normal, degraded, and emergency modes. Information delivered acoustically should be audible to those who need to receive it within the expected acoustic environment.

Alarms and alerts

- Alarms and alerts should be designed with consideration of criticality, levels of alarms, and the role of end users.
- In general, alarms indicate a condition that requires action, while alerts provide information about performance degradation that may affect expected functionality.
- Critical alarms should be distinguishable by the end user and notified in both audible and visual forms.
- Alarms should be prioritized based on their importance, typically limited to no more than three levels of alarms.
- It should be ensured that end users can hear audible alarms within the expected acoustic environment.
- In computer systems where several different users access the system with separate role signs-ins, alarms and alerts should be designed to be delivered only to the relevant user in the most appropriate form for that role. For example, in a train operating system that allows drivers, guards, and maintainers to sign in as different roles, then the alarms and alerts for drivers should differ from those for guards and maintainers.

Control room

To achieve the expected performance and prevent operator errors, HF should be considered in control room design.

- Control room equipment should be arranged based on the operators' functions.
- An analysis should be conducted to understand the required interactions between operators in the control room and between operator and equipment, aiming to reduce movements across the room where the operator may have to leave one console to attend another.
- The operator's usual seated position should be near the most frequently used console.
- The design should ensure that in any seated position, the operator faces the display board.
- Visibility should not be obstructed by any console or person.
- Pathways should be designed to restrict non-essential personnel from the main working areas.

In the design of consoles, the console height should be calculated so that even the shortest operator can see remote monitor walls or displays over any mounted electronics.
Since control room operators rely heavily on their vision to monitor screens, displays, and other digital devices, it is essential to create condition that minimise eye strain. This can be achieved by placing items and screens at similar distances, positioning displays at an appropriate angle to the line of sight, increasing task lighting on printed material, and using larger text sizes. Lowering monitor height is also recommended, as this can help reduce eye strain and decrease the risk of dry eye syndrome, which occurs when eyelids open wider while looking upwards.

Optimal viewing angles for an operator involve minimal eye movement and no head movement, especially for essential and frequently performed tasks. The physical layout should also account for non-electronic equipment and documents, with flexible positioning for items like telephones, keyboards, mice, controllers radio/intercom and writing areas. This flexibility allows operators to adjust during shifts, helping to reduce fatigue.

Other considerations

< Seating >
Seating plays a critical role in ensuring suitability and appropriate dimensions within Human Factors. Designers and suppliers of equipment and facilities should select, and design seating and its layout based

on anthropometric data for the end user. They must also justify any compromises made due to access and egress or specific end user requirements regarding seating and standing capacity or other conflicting needs.

< Glare and reflection >

Human Factors requires elimination and minimisation of glare and reflection. Designers should eliminate glare and reflections wherever possible. Where elimination is not feasible, glare and reflections should be minimised on workstations, controls, and displays, with particular attention to control rooms.

< Customers and the public >

Human Factors also considers customers and the public to ensure no discrimination against users' physical disabilities, improved information provision for customer and the public, more convenient customer seating, and the appropriate placement of handrails, poles, and grab points.

< Users with disabilities >

Designers should design and provide equipment and facilities considering users with disabilities to remove any barriers in all circumstances. These should comply with relevant regulations and legislation in each country, with conformance verified by the client and assessed by a HF specialist.

< Information for customers and the public >

Designers should ensure customer and public information is clear, using optimized language, clarity, symbols, and testing. Information should be clear and unambiguous enough to guide customers or the public effectively and safely within the transport system, especially to take appropriate actions during abnormal or emergency event. Symbols should be used where applicable to convey information.

HF Validation

HF Validation involves assessing detailed designs against the HF criteria to ensure they meet the requirements and follow good practice.

Human Factors assessment

Human factors assessment refers to the systematic evaluation of whether design aspects meet Human Factors (HF) requirements at the end of the design phase. The following processes outline the steps in HF assessment:

- Prepare HF validation plan.
- Validate the detailed design against the HF criteria, which may include 3D human CAD assessment.
- Collect and review designers' and suppliers' validation information.
- Update the Human Factors Integration log to reflect findings.
- Hold workshops with designers and suppliers to resolve outstanding HF issues through a collaborative approach.

Adopt and test

Human factors adoption and testing involve integrating human factors principles into the design and evaluation of products, systems, or processes, followed by rigorous testing to assess their usability and effectiveness. The significance of this process lies in its ability to ensure that designs meet the needs and capabilities of users, enhancing safety, efficiency, and satisfaction. By adopting human factors principles, organisations can reduce the risk of errors, improve user experience, and ultimately create solutions that are more intuitive and accessible. Testing these designs helps identify any remaining issues, allowing for adjustments before full implementation, which is essential for optimising overall performance.

HF issues and requirements can be addressed as a part of safety issues and requirements, meaning that all HF issues (or hazards) should be input into the safety hazard log database and managed accordingly.

INTERFACE MANAGEMENT

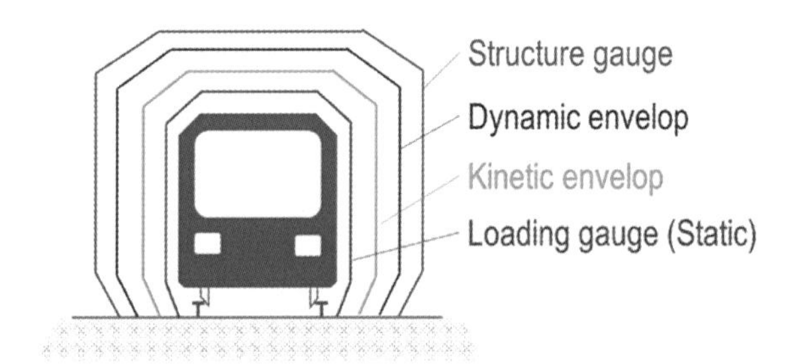

Overview of interface management

Since a system is an integrated set of subsystems that interact with each other and the environment, it is necessary to manage interface issues between them. Moreover, each subsystem is generally developed by different contractors, and design of a system can be influenced by multiple stakeholders. Additionally, the system may interact with other external systems.

Therefore, an interface can be defined as follows:

- the physical or functional boundary between two or more adjacent elements
- any scope of work shared by more than one party (or discipline)
- any interface or disputed items related to subsystems or organisation (or stakeholders).

Objectives of interface management

The objectives of interface management are to:

- identify, clarify,
- resolve, control, and
- document all interfaces and interferences between system components, as well as any disputes between subcontractors.

In interface management, identifying and clarifying interface items is crucial for several reasons:

- Clear communication: Proper identification ensures that all stakeholders understand the specific interfaces involved, reducing confusion and miscommunication.
- Risk management: Clarifying interface items helps identify potential risks and dependencies early in the project, allowing teams to address issues proactively.
- Responsibility assignment: Clearly defined interfaces enable better assignment of responsibilities among teams (or subcontractors), ensuring everyone understands their roles regarding the interfaces.
- Efficient integration: Well-defined and understood interfaces lead to smoother integration of components, reducing delays and enhancing overall project efficiency.

Documenting interface activities and outcomes in interface management is important for several reasons:

- Traceability: Documentation provides a clear record of actions taken, allowing teams (or subcontractors) to trace decisions and changes throughout the project.
- Knowledge sharing: Well-documented activities and results facilitate knowledge transfer among team members and across projects, helping new members understand past actions and decisions.
- Quality assurance: By documenting processes and outcomes, teams can assess performance and ensure that interfaces meet quality standards, reducing the risk of errors.
- Dispute resolution: In case of disagreements or misunderstandings, documented evidence can help clarify responsibilities and decisions, supporting conflict resolution.
- Future reference: Documentation serves as a valuable resource for future projects, offering insights and lessons learned to enhance efficiency and effectiveness.

Roles of Interface Manager

An interface manager should address both interface and interference issues. However, it is not necessary for an interface manager to manage interface issues within a subcontractor's internal operations. The ultimate role of an interface manager at the top level of the organisational structure is to manage responsibilities between subcontractors. In this chapter, "a subsystem" refers to the scope of a subcontractor's contract. The role of an interface manager is critical in project management, particularly in complex projects involving multiple teams (or subcontractors) or components. Key aspects of their role and importance include:

- Coordination: The interface manager coordinates communication and collaboration between different teams, ensuring alignment on project goals and interfaces.
- Issue resolution: They proactively identify and resolve interface-related issues, minimising disruptions and maintaining project timelines.
- Documentation: The interface manager oversees the documentation of interface requirements, changes, and activities, providing a clear reference for all stakeholders.
- Risk management: By monitoring interfaces, they identify potential risks early and implement strategies to mitigate them, ensuring smoother project execution.
- Quality Control: The interface manager ensures that all interfaces meet established quality standards, contributing to the overall success of the project.

Interface or interference issues

Types of interface or interference issues in railway construction projects can be classified as follows:

- Data flow / power transmission – between subsystems [e.g., input and output of signalling & communication systems]
- Spatial interference – between subsystems or between system and structure [e.g., structure gauge]
- Physical contact – along a line [e.g., station and civil work]
- Missing item – items that need to be designed but have not been identified
- Design considerations – between subsystems [e.g., tunnel cross-sectional area and rolling stock cross-sectional area]
- Responsibilities – allocation of RAM target, safety requirements, etc.
- Time interface – between predecessors' and successors' activities [e.g., construction sequences]

Data flow and power transmission

To perform the required functions, subsystems continuously interact with each other. For instance, wayside signalling and onboard signalling systems communicate with each other to control rolling stock. Effective data flow ensures that real-time information is shared accurately among trains, the control centre, and operation teams, enabling efficient operations, safety monitoring, and timely decision-making. In this case, at least two factors need to be coordinated: communication protocol and the voltage.

Figure 71 – ETCS

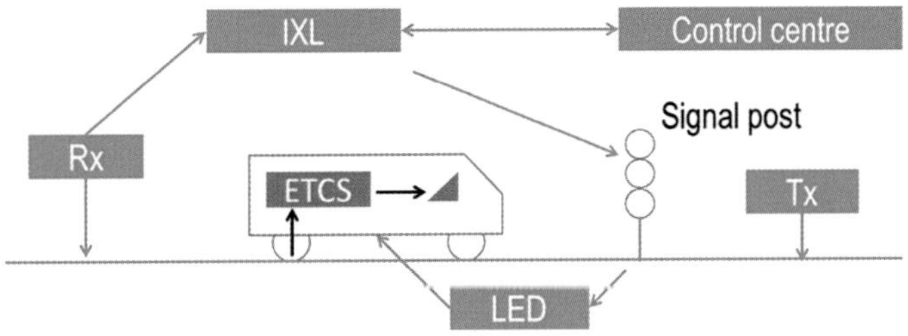

Power transmission involves the distribution of electrical power from the source (such as overhead lines or substations) to train systems, including traction systems and onboard equipment.

Design consideration

Here are some examples of design considerations in interface management between railway subsystems:

- Tunnel design: When a train passes through a tunnel, the tunnel's cross-sectional area must be appropriately designed to minimise the air compression effects caused by the moving train. This requires coordination between tunnel dimensions and the cross-sectional area of the rolling stock.
- Signalling coordination: The interface between signalling systems and train control systems needs careful consideration to ensure accurate monitoring and control of train movements, preventing collisions and ensuring safety.
- Power supply interfaces: The interface between overhead catenary systems and train power systems must be designed to ensure reliable power transfer, considering factors such as voltage levels, electrical frequency, current capacity, and connector compatibility.
- Noise and vibration management: When designing interfaces between tracks and rolling stock, considerations must be made to minimise noise and vibration impacts on surrounding communities. This requires collaboration between civil engineering team and rolling stock team.
- Maintenance accessibility: Interfaces between various subsystems (e.g., track, signalling, and train maintenance) should be designed to allow easy access for maintenance and inspections, facilitating efficient upkeep and minimising downtime.
- Emergency response protocols: Communication interfaces between different subsystems, such as train control and emergency services, must be clearly defined and tested to ensure rapid response in case of incidents.

The design considerations help to ensure safety, efficiency, and effectiveness across various railway subsystems.

Spatial interference

One of the most important spatial interferences in railway infrastructure is the structure gauge. The structure gauge is the minimum clearance outline that defines the limits of railway infrastructure to prevent interference with the movement of rolling stock. There are two types of gauges – loading gauge and structure gauge in railway infrastructure, as shown in Figure 72.

Figure 72 – Gauges and envelopes

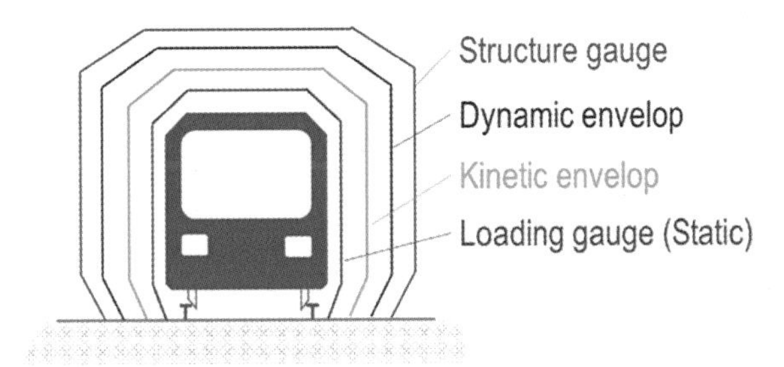

- Loading gauge – the maximum permitted cross-sectional dimension of a vehicle. The shape of the vehicle and any equipment on it must not protrude outside of the gauge.
- Kinetic envelope – the outline that allows for the maximum movement (due to conicity of wheels, where a train moves in a snake-like motion, known as "hunting," as shown in Figure 73) of a travelling vehicle. To check the envelope, rolling stock engineers and rail engineers must collaborate.

Figure 73 - Hunting movement

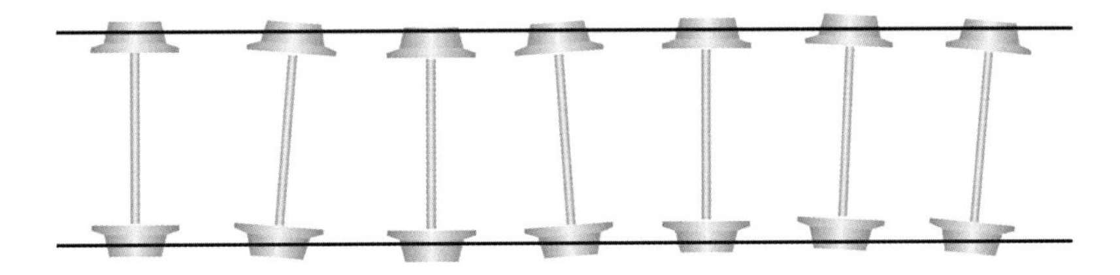

- Dynamic envelope – the space occupied by the end protrusions and the centre protrusion of rolling stock travelling along a curve, as shown in Figure 74. The dynamic envelope is set outside of kinetic envelope, and structure gauge must be set outside the dynamic envelope. Additionally, the shape of the infrastructure and any infrastructure equipment must be installed outside the structure gauge.

Figure 74 - Central and end protrusions

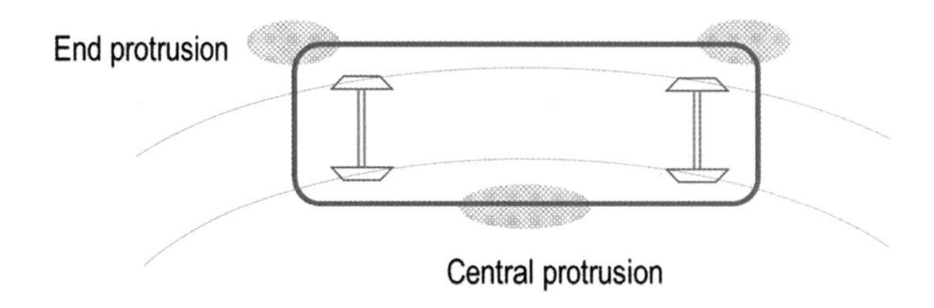

Physical contact

There are many physical contacts between subsystems (or subcontractors' deliverables). Here are some examples of physical contact considerations in interface management between railway subsystems:

- Track and substructure: The installation of tracks on subgrade structures must ensure proper alignment and support to prevent track deformation, requiring careful coordination between civil engineering and track laying teams.
- Signalling systems on tracks: When signalling systems are installed along the tracks, they must be physically aligned and securely mounted to ensure accurate signal functionality.
- Station interfaces: The junction between mainline tracks and station platforms requires precise design to facilitate safe and efficient boarding and alighting of passengers, including considerations for gaps between the train body and the platform edge.
- Overhead catenary system: The physical contact between the overhead catenary system and the pantographs on trains must be designed to ensure reliable power transfer while accommodating movements and fluctuations during train operation.

The physical contact considerations are essential for maintaining safety and efficiency in railway operations.

Responsibility allocation

Functional requirements at highest level, safety requirements from PHA (Preliminary Hazard Analysis), and the overall RAM targets should be broken down and allocated to each subcontractor. For RAM target allocation, refer to the chapter "RAM allocation".

When it comes to safety requirements, safety-related responsibilities can be assigned based on outputs from Preliminary Hazard Analysis. Each subcontractor, such as those involved in signalling, track design, and rolling stock, is allocated specific safety requirements that they must address within their systems.

Missing item

When dividing the entire scope and allocating it to subcontractors, it is not possible to perfectly allocate the full scope of work without missing items. During this process, some important items may be overlooked, and these missing items can cause project delay due to re-design or re-work when they are identified later. Therefore, missing items should be identified in advance to prevent such issues.
The interface manager can encourage subcontractors to identify missing items from the early stages. The requirements manager can also detect missing items during the decomposition process.

Time interference

The shape of railway construction site is generally narrow, which means that two or more disciplines cannot perform their tasks simultaneously on the same site. If schedule management fails, time interference can occur between predecessors and successors. A schedule controller can identify such interference by grouping activities that are supposed to be performed at the same time and in the same location using schedule management software.
The following steps outline how to identify interference using Primavera P6:

- Identify the areas and spaces that are expected to be crowded by different subcontractors.
- Create activity codes and assign them to activities according to identified locations.
- Reorganise activities based on activity codes using "Group and Sort" feature. For example, Figure 75 shows activities to be performed on 3rd floor.
- Review Gantt charts for each area and space. If there are locations where multiple tasks will be conducted simultaneously and expected to be crowded with different subcontractors, these should be identified and resolved. (The red ellipse highlights the collection of tasks to be performed at the third floor in May 2017.)

Figure 75 – Grouping by location

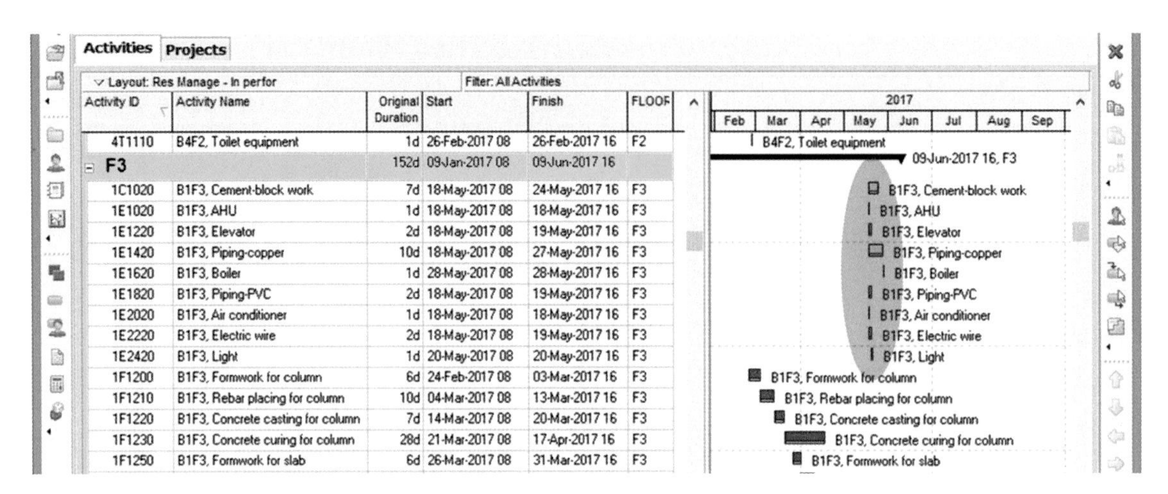

Activity ID	Activity Name	Original Duration	Start	Finish	FLOOR
4T1110	B4F2, Toilet equipment	1d	26-Feb-2017 08	26-Feb-2017 16	F2
F3		152d	09-Jan-2017 08	09-Jun-2017 16	
1C1020	B1F3, Cement-block work	7d	18-May-2017 08	24-May-2017 16	F3
1E1020	B1F3, AHU	1d	18-May-2017 08	18-May-2017 16	F3
1E1220	B1F3, Elevator	2d	18-May-2017 08	19-May-2017 16	F3
1E1420	B1F3, Piping-copper	10d	18-May-2017 08	27-May-2017 16	F3
1E1620	B1F3, Boiler	1d	28-May-2017 08	28-May-2017 16	F3
1E1820	B1F3, Piping-PVC	2d	18-May-2017 08	19-May-2017 16	F3
1E2020	B1F3, Air conditioner	1d	18-May-2017 08	18-May-2017 16	F3
1E2220	B1F3, Electric wire	2d	18-May-2017 08	19-May-2017 16	F3
1E2420	B1F3, Light	1d	20-May-2017 08	20-May-2017 16	F3
1F1200	B1F3, Formwork for column	6d	24-Feb-2017 08	03-Mar-2017 16	F3
1F1210	B1F3, Rebar placing for column	10d	04-Mar-2017 08	13-Mar-2017 16	F3
1F1220	B1F3, Concrete casting for column	7d	14-Mar-2017 08	20-Mar-2017 16	F3
1F1230	B1F3, Concrete curing for column	28d	21-Mar-2017 08	17-Apr-2017 16	F3
1F1250	B1F3, Formwork for slab	6d	26-Mar-2017 08	31-Mar-2017 16	F3

Interface techniques and tools

Identification of interfaces

Identifying interface and interference issues is a critical activity because it allows relevant teams to begin their interface task with clearly identified interfaces. The following list provides sources where interfaces are initially identified:

- Contractual documents: Some contractual documents may include interface items or interference issues.
- Interface matrix: To create an interface matrix, interactions between different components are developed in a grid-like format with both rows and columns representing sets of components. This matrix helps visualise and manage complex interactions.
- Historical interface items and issues from similar projects: This is one of the best and easiest ways to identify interface items and issues.
- Interface items from designers: During design process, each designer requires input data from other disciplines and is aware of the necessary design considerations. Their insights can be utilised to identify interface items.
- Document review: Interface managers should continuously review design documents to identify design considerations, missing items, and interface issues.
- Interface items published on websites: Interface items can be found on websites through online search.

Interface issues and interface issue register

An interface issue is an interface that is not fully defined and understood, and for which the responsibility for management and delivery has not been agreed upon. As a result, the identified interface issue cannot be resolved quickly. Any identified interface issue should be recorded on a log sheet – the interface issue register. The interface issue register will be used to define the responsibility and delivery of identified interface issues through interface meetings. To effectively manage the identified issues, the register database should include the columns or fields:

- Interface issue ID: A unique number
- Originator: The team or person who raised the interface issue

- Owner organisation: The organisation to which the interface issue owner belongs
- Affected parties: A list of all internal and external parties with interests in the interface issue
- Interface issue detail: A description of the issue
- Raised date: The date when the interface item is raised
- Safety relevant: Indicates whether the issue has safety implications (a safety manager should use this information when conducting Interface Hazard Analysis)
- Completion date: The date when the issue is resolved or completed
- Status: The status of the issue – open, closed, removed, etc.
- Interface issue priority: Categorisation of the issue in terms of critically, based on the level of urgency

Master interface database

The master interface database (or worksheet) is a collection of interface information where all the closed interface items and issues are stored, along with supporting evidence. It is used for reference and to track the results of interface activities. The evidence stored in the database will be crucial for resolving disputes between parties (or subcontractors) and teams regarding interface issues raised during the T&C (Testing and Commissioning) phase.

The database should have fields or columns as follows:

- Interface ID: A unique number
- Reference ID: The ID number from the Interface Issue Register
- Owner: The contributor responsible for providing information such as plans, drawings, specification, calculations, testing results, etc.
- Interface partner: The party that who requires the information provided by the owner
- Detail: A description of the interface
- Safety relevant: Whether the interface is related to safety
- ICD ID: If the interface item is related to ICD (Interface Control Document), it should have ICD ID number
- Close date: The date the issue is closed
- Evidence: The location where the evidence is stored

Interface management process

Figure 76 outlines the interface management process.

Figure 76 - Interface management process

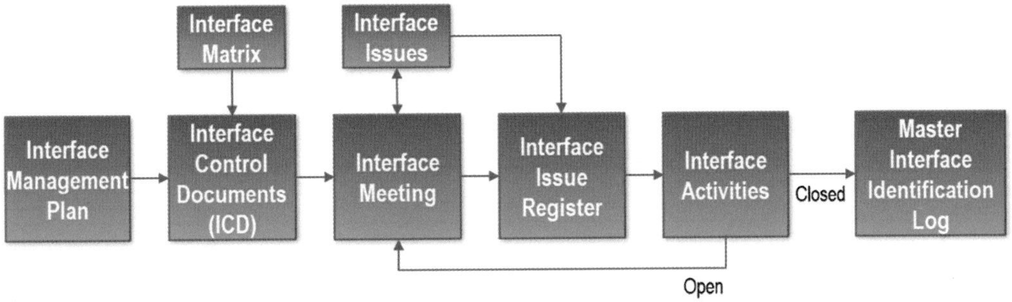

- Develop the interface management plan and interface matrix.
- Create the ICD (Interface Control Document) based on the interface matrix.
- Gather the interface items and issues.
- Hold interface meetings.
- Register the issues in the interface issue register.
- Take actions to resolve the interface issues, such as providing documents, meeting, etc.
- Regularly maintain the interface issue register.
- Record closed interface issues in the master interface identification log.
- Verify and validate interface issues at each phase.

Interface Management Plan

Unlike other plans, the interface management plan should be shared with stakeholders to ensure their involvement. Therefore, an Interface Management Plan should:

- Define a procedure for involving the stakeholders
- Clearly define the scope of the interface management
- Identify relevant stakeholders

- Include the interface management process
- Outline the process for interface activities

Since the purpose of the interface management plan is to encourage active participation from relevant teams in interface activities, the plan should be developed in a clear and understandable manner, using graphical process, clear terminology, and directions. The interface management plan should include the following contents:

- Interface identification methodologies
- Control of interface issues
- Interface management process
- Roles and responsibilities

Interface Matrix

The interface matrix defines the relationships between subsystems. It can be developed by data from similar projects and serves as foundation for developing ICD (Interface Control Document). To create the interface matrix, all subsystems should be broken down, with the level of breakdown depending on the organisational structure of the project.

Figure 77 - An example of interface matrix

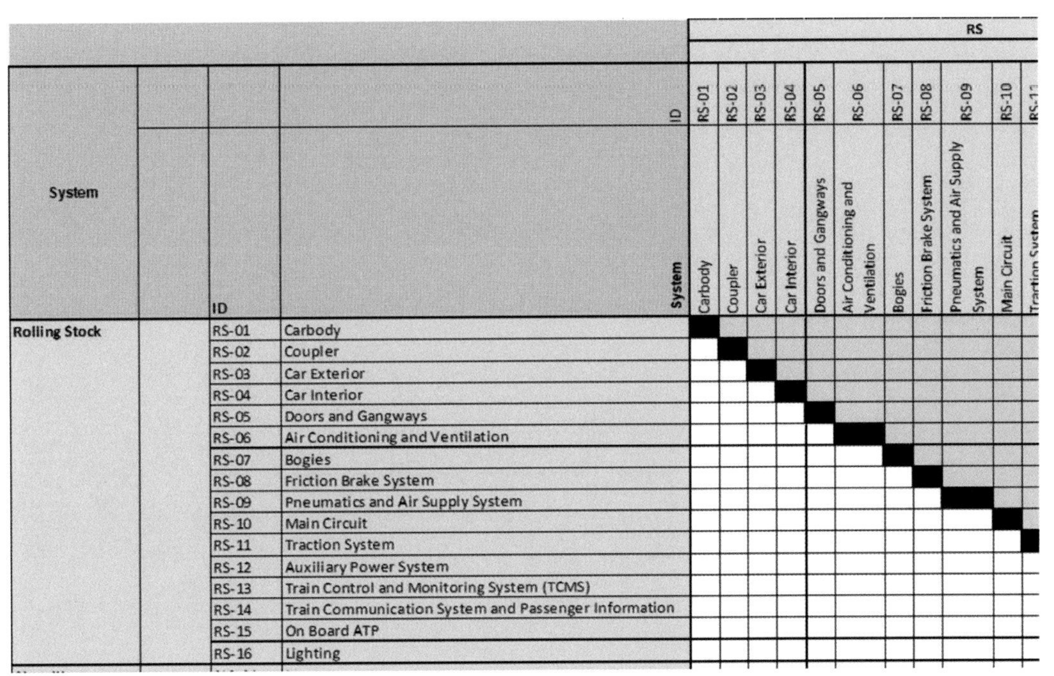

System		ID	System / ID	RS-01 Carbody	RS-02 Coupler	RS-03 Car Exterior	RS-04 Car Interior	RS-05 Doors and Gangways	RS-06 Air Conditioning and Ventilation	RS-07 Bogies	RS-08 Friction Brake System	RS-09 Pneumatics and Air Supply System	RS-10 Main Circuit	RS-11 Traction System
Rolling Stock		RS-01	Carbody											
		RS-02	Coupler											
		RS-03	Car Exterior											
		RS-04	Car Interior											
		RS-05	Doors and Gangways											
		RS-06	Air Conditioning and Ventilation											
		RS-07	Bogies											
		RS-08	Friction Brake System											
		RS-09	Pneumatics and Air Supply System											
		RS-10	Main Circuit											
		RS-11	Traction System											
		RS-12	Auxiliary Power System											
		RS-13	Train Control and Monitoring System (TCMS)											
		RS-14	Train Communication System and Passenger Information											
		RS-15	On Board ATP											
		RS-16	Lighting											

There are two types of interface matrices as follows:

- N1 interface matrix: This matrix maps the interactions and relationships between different components, as shown in Figure 78 (left-hand side).
- N2 interface matrix: This matrix not only shows the connections between components but also illustrates the direction of data flow between them, as shown in Figure 78 (right-hand side).

__Figure 78 - N1 and N2 interface matrics__

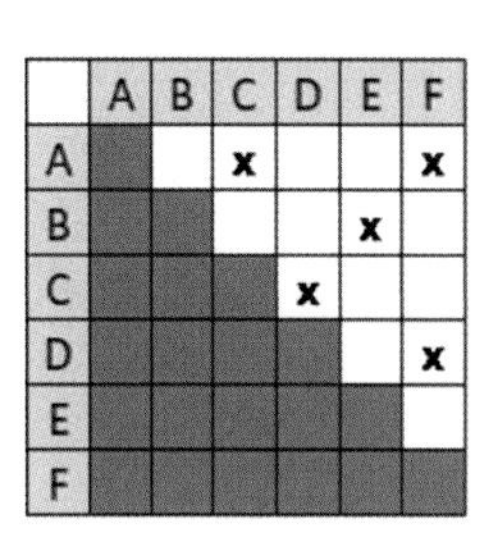

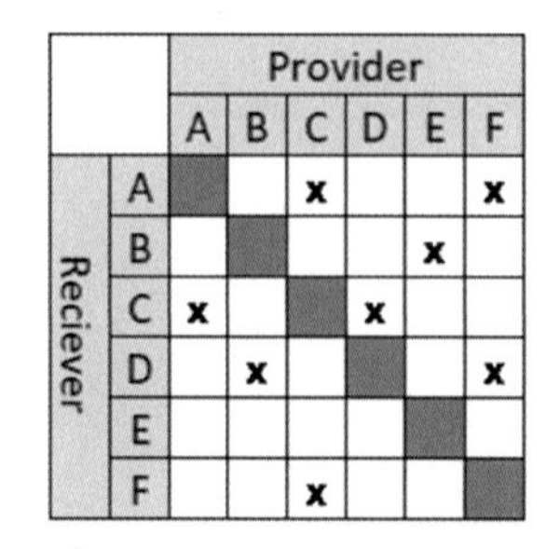

Interface Control Document (ICD)

An Interface Control Document (ICD) is a technical document that outlines the details of interactions between two or more systems, subsystems, or components. It specifies how these systems communicate, including the format, protocols, data types, and timing of the exchanges. ICDs are used to ensure effective collaboration between different subsystems by clearly defining the requirements for their interfaces, helping to prevent integration issues.

ICD (Interface Control Document) should assist in defining interface issues, as well as the inputs and outputs between components to be discussed. It should also help in identifying stakeholders and their respective roles and responsibilities.

An ICD generally consists of the following:

- Overall data flow: Shows the overall direction of data flow between components.
- Interface description: Details the data type, direction of data flow, and any relevant specifications.
- Roles and responsibility: Outlines the roles and responsibilities of each party and their activities at each phase.
- Verification and validation (V&V): Provides information on interface requirements, results of V&V, and supporting evidence

Figure 79 – An example 3of ICD

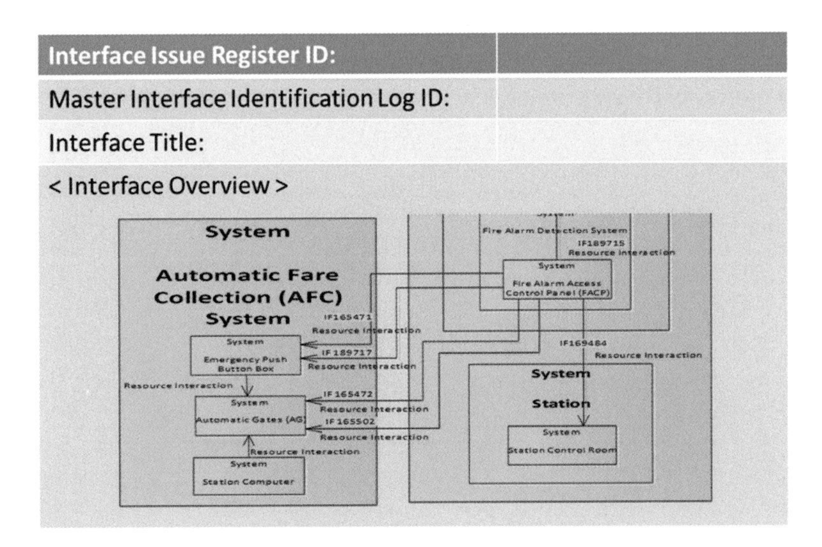

Interface meeting

An interface meeting is organised to discuss interface issues, where new interface issues can be raised. Before holding the meeting, the interface manager should gather the agenda and distribute it to relevant teams (or subcontractors).

Here are some effective tips for conducting interface meetings to resolve interface issues:

- Define clear objectives: Set a clear agenda with specific goals for the meeting. Ensure all participants understand the purpose and desired outcomes.
- Invite relevant stakeholders: Include representatives from all affected teams or subcontractors to ensure that diverse perspectives are considered and that relevant issues are addressed.
- Prepare thoroughly: Distribute meeting materials in advance, including relevant documents, data, and previous meeting notes. This allows participants to come prepared to discuss the issues.
- Facilitate open communication: Encourage an open and respectful dialogue, where all participants feel comfortable sharing their insights, concerns, and suggestions.
- Identify key issues: Focus on the most critical interface issues first. Use structured techniques, such as root cause analysis, to explore problems and identify underlying causes.
- Document discussions: Keep detailed minutes of the meeting, including decisions made, action items, and responsible parties. This documentation serves as a reference for follow-up.

- Assign action items: Clearly assign tasks to individuals or teams, specifying deadlines for resolution. Ensure accountability by tracking progress on these action items in future meetings.
- Follow up: Schedule follow-up meetings to review progress on action items and continue addressing any outstanding issues. Regular check-ins help maintain momentum and accountability.
- Evaluate effectiveness: After the meeting, gather feedback from participants to assess the effectiveness of the meeting format and adjust as needed for future meetings.
- Foster a collaborative culture: Promote a culture of collaboration and problem-solving, encouraging teams to work together to find solutions to interface issues rather than focusing on blame.

By implementing these practices, interface meetings can become more efficient and productive, ultimately leading to fast and better resolution of interface issues.

< Interface activities >

All communications related to solving interface issues should be traceable. When interface problems arise, it is essential to define the responsible subcontractors or teams. Therefore, all communication in interface activities should be documented using official transmittals, letters, memos, etc.

If an interface issue is resolved, it should be logged in the Master Interface Identification Log along with supporting evidence. Engineers should continue to hold interface meetings and conduct necessary activities until all interface issues are resolved.

CONFIGURATION MANAGEMENT

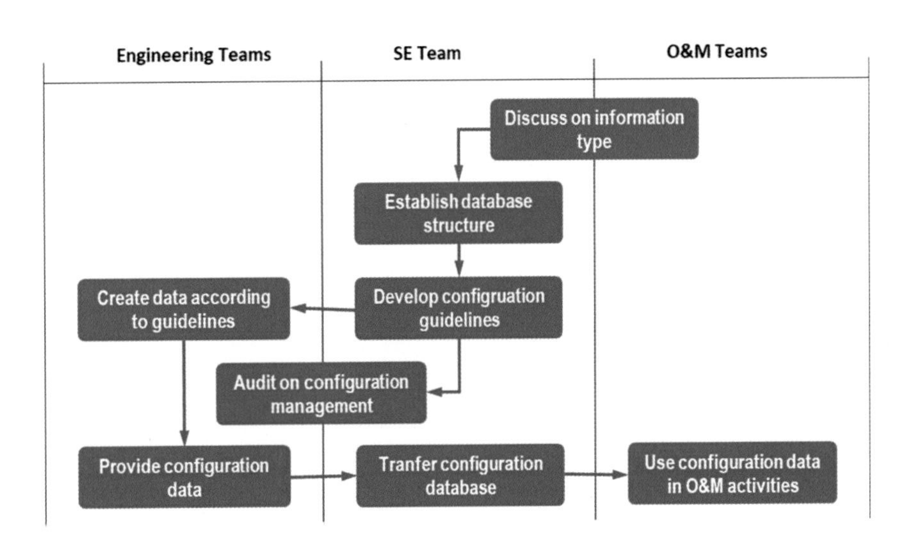

Overview of configuration management

Configuration management (CM) ensures that a system remains consistent, reliable, and adaptable to changing requirements throughout its lifecycle. In Systems Engineering, CM aims to identify system components and control changes during the development phase. The CM database, generated from CM activities, serves as a configuration database for maintenance during the operation phase.
CM data typically includes physical deliverables, their components, and relevant documents such as drawings, design reports, calculation sheets, and specifications. Since a railway project is generally a large project consisting of many components, one important activity of CM in railway projects is aligning informational documents, such as drawings, with physical deliverables to help operators and maintainers manage the railway infrastructure efficiently.

Advantages of Configuration Management (CM)

Configuration management provides a structured and systematic approach to managing the complexity of product or system development and maintenance, ensuring that changes are well managed, consistent, and traceable.
The specific advantages of configuration management include:

- Identifying and controlling changes: Tracking and controlling changes to the product or system, including documenting the history of changes.
- Ensuring consistency and traceability: Maintaining consistent configurations throughout the development and maintenance of a product or system.
- Improving collaboration and coordination: Facilitating collaboration and coordination among stakeholders by providing a clear and up-to-date understanding of the configuration.
- Facilitating testing and evaluation: Enabling the testing and evaluation of individual components and the entire system to ensure they meet requirements and expectations.
- Supporting problem resolution: Providing information and tools to support problem resolution, reducing the time and effort required to address issues.
- Ensuring compliance: Ensuring compliance with relevant standards, regulations, and specifications.
- Facilitating knowledge transfer: Facilitating the transfer of knowledge from one project phase to another, or from one project to another.

The configuration database generated by configuration management activities can be used as the database for the asset management system and maintenance management system during operation and maintenance phase.

Process of CM and guidelines

To manage configuration items effectively, configuration management should be well-defined and consistent, helping an organisation improve the quality and reliability of their products and systems. The configuration management process includes the following steps:

- Develop a Configuration Management Plan: Defining the objectives, processes, and procedures for configuration management.
- Establish a Configuration Control Board: Establishing a board responsible for approving changes to the configuration and ensuring that change control procedures are followed.
- Implement version control: Implementing a version control system to track changes to the configuration and ensure that only approved changes are made.
- Identify configuration items: Identifying all the configuration items that need to be managed, including hardware, software, documents, and other artifacts.
- Establish a baseline: Establishing a baseline configuration that defines the starting point for the development or maintenance of the product or system.
- Implement change control: Implementing a change control process that ensures that changes are properly documented, reviewed, approved, and implemented.
- Conduct configuration audits: Conducting regular configuration audits to ensure that the configuration management processes are being followed and that the configuration is consistent with the baseline.

Figure 80 illustrates a summary of the configuration management process, where the roles of O&M teams are to provide information types to the SE team, the roles of engineering teams are to create and provide configuration data, and the roles of the SE team are to manage the overall CM process.

Figure 80 - Summary of configuration management process

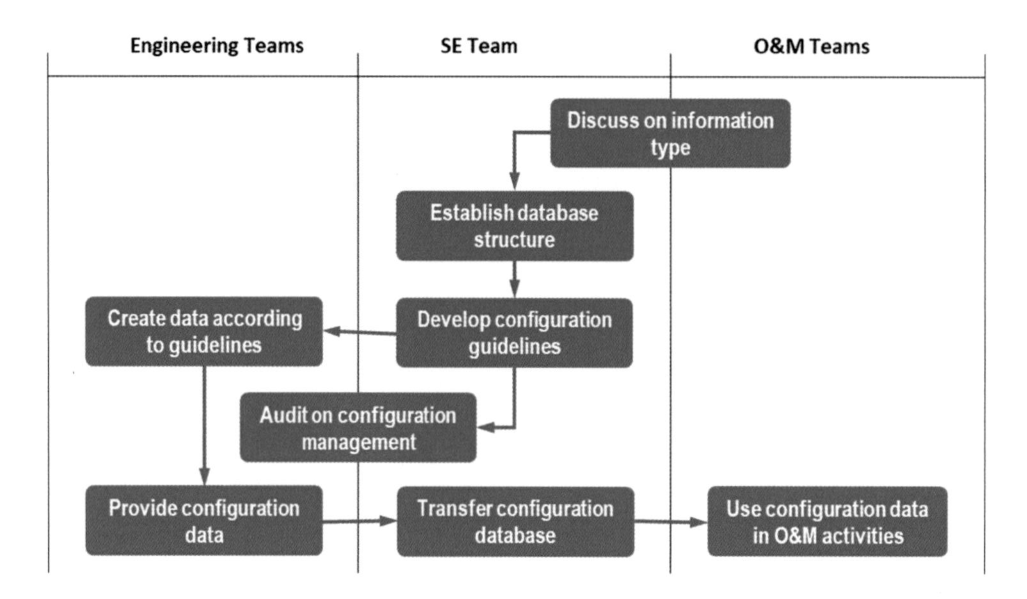

Configuration Management Plan

A Configuration Management Plan describes how to manage the configuration of a project. In railway projects, the focus is primarily on E&M (Electric and Mechanical) systems.

Scope of work and strategies

In railway projects, target systems for configuration management can include the following:

- Signalling and train control
- Traction power supply
- Communication (including AFC, SCADA)
- Depot equipment
- Building management service (BMS)
- Platform screen door (PSD)

Since there are many configuration items in railway projects, configuration management strategies should be established carefully. When establishing the strategies:

- Operation and maintenance should be a focus.
- An asset-based approach should be considered.
- The system configuration hierarchy should be clear.
- The configuration management and change management processes should not be overly complex to avoid construction delay.

Here are some references for configuration management plan:

- ISO 10007: Quality management – Guidelines for configuration management
- ISO 9001: Quality management systems – Requirements
- BS/ISO 10007: Quality management systems – Guidelines for configuration management

CM activities in each phase

A Configuration Management Plan must include configuration management activities for each phase to ensure proper control, traceability, and consistency of the project's configuration items throughout the entire lifecycle. This ensures that any changes are systematically managed, the system remains aligned with its specifications, and all stakeholders are informed and synchronised, reducing risks and maintaining quality and compliance.

- Detailed design phase: Developing a Configuration Management Plan, including the change control process, identifying configuration items, maintaining configuration items, and establishing the baseline for the detailed design phase.
- Manufacturing & installation phase: Controlling configuration changes, identifying additional items, maintaining configuration items, performing configuration audits, and establishing the baseline for the manufacturing and installation phase.
- Testing & commissioning (T&C) phase: Controlling configuration changes, identifying additional items, maintaining configuration items, performing configuration audits, establishing the baseline for the T&C phase, and handing over the configuration database to the project owner (or to the asset manager and maintenance manager).

The CM process in Systems Engineering typically involves the following steps:

- Configuration identification: The process of identifying the components of a system that need to be managed and the criteria for determining their status as configuration items (CIs).
- Configuration control: The process of controlling changes to CIs and ensuring that changes are made in accordance with established procedures and processes.
- Configuration status accounting: The process of recording and reporting the status of CIs, including changes made and the impact of those changes on the system.
- Configuration verification and audit: The process of verifying that the system adheres to established configuration requirements and standards, and that changes are properly recorded and reported.
- Configuration management documentation: The documentation that supports the CM process, including policies, procedures, and work instructions.

Roles and responsibilities

A Configuration Control Board (CCB) is a group of stakeholders responsible for managing changes to a project's scope or technical specifications. The CCB is typically led by the client, as most major changes involve cost and/or schedule adjustments that require the client's approval, as shown in Figure 81.
The main roles and responsibilities of CCB are as follows:

- A responsible working group for all the project configuration management activities for the system

- Reviewing and approving identified configuration items
- Reviewing and approving the establishment of baselines
- Reviewing and approving proposed configuration changes
- Conducting other configuration management activities as required by the project contract terms and project environment conditions

Figure 81 – Process of configuration changes

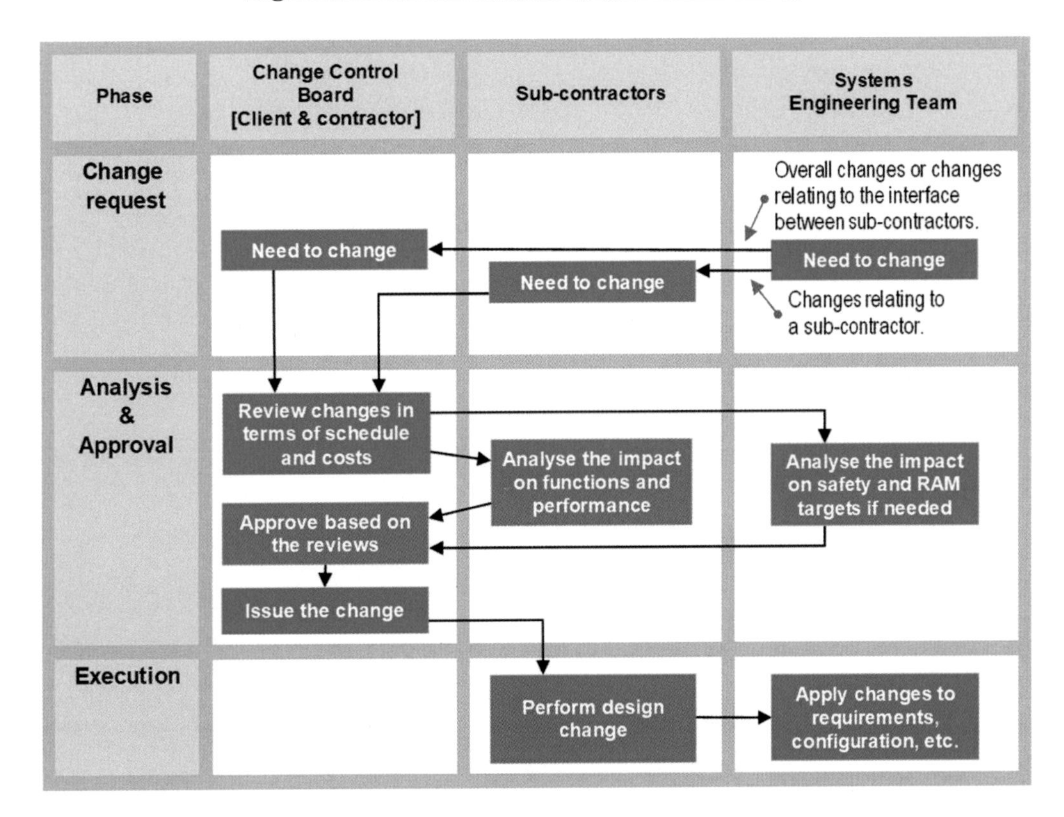

A configuration manager works closely with other stakeholders to ensure that all configuration management systems are aligned with the project's goals, requirements, and standards. The main roles and responsibilities of configuration managers are as follows:

- Responsible for the Project Configuration Management Plan
- Implementation of the configuration management process at the project level

- Configuration accounting and its maintenance
- Baseline establishment
- Configuration audit
- Coordination with the CCB

Configuration managers in subcontractors are staff who work at the subsystem level (depending on the contract terms) for configuration of subsystems, as follows:

- Responsible for the implementation of the configuration management process at the subsystem level
- Configuration item identification
- Subsystem breakdown and preparation of configuration worksheets
- Documentation of design, installation, and testing, including drawings
- Submitting configuration change request to the CCB
- Preparation of Change Request Form (CRF) with relevant evidence, including impact analysis in terms of cost, schedule, and interface
- Maintenance of configuration sheets and periodical reporting to the configuration manager for current configuration accounting
- Implementation of the approved Configuration Change Order (CRO) from the client
- Support of configuration audits and corrective actions based on audit findings
- Participation in CCB meeting if required

Identification, control and status accounting

Configuration identification

Configuration identification is a process in Configuration Management (CM) that involves identifying and documenting the specific components or elements that make up a product, system, or project. This process helps define the product's technical baseline, which serves as a reference for managing changes throughout the product's lifecycle.
The configuration identification process involves creating a bill of materials (BOM) or inventory of all the components that make up the product or system and assigning unique identification numbers or labels to each component. This information is used to track changes to the components, including updates, modifications, and replacements. The process also includes creating and maintaining configuration documentation, such as technical specifications, design documents, and test plans, which provides a clear and complete picture of the product's current state.

< Numbering structure >
The numbering structure for each configuration item (CI) in Configuration Management (CM) is a system for assigning unique identification numbers or labels to each CI. The numbering structure provides a standardised and consistent method of identifying and tracking CIs throughout their lifecycle.
The numbering structure for a CI can vary depending on the project scale and specific requirements, but it typically includes a combination of letters and numbers that represent different aspects of the CI, such as its type, location, and version. For instance, a numbering structure for a railway system might include letters to represent the type of CI (e.g., T for track, S for signal, R for rolling stock), and numbers to represent the order of the CI.
The numbering structure for CIs should be clear, concise, and easily recognisable. It should also be flexible enough to accommodate changes and growth in the number of CIs over time. The numbering structure must be consistent across the entire project and included in all CI documentation, such as the bill of materials (BOM) or inventory, and configuration documentation. This ensures that all stakeholders are aware of the numbering structure and can easily track and identify CIs.

Configuration change classes and change control process

Configuration changes can be categorised into two categories as follows:

- Class 1: Changes that directly impact the terms or conditions of the project contract, such as project scope, product functions, cost, schedule, RAM target, and baselines.
- Class 2: All other changes that do not belong to Class 1, such as correction of typo in documents, minor errors, etc.

Even though the configuration control process typically involves a Configuration Control Board (CCB) responsible for evaluating and approving changes to the CIs, the configuration changes go through the CCB depending on their class. Class 1 changes should be officially requested to the CCB, while Class 2 changes may be reviewed and implemented by individual configuration managers of subcontractors without requesting the CCB. However, Class 2 changes should be recorded in the configuration status accounting to ensure clear traceability.

Major records under configuration change control are as follows:

- Configuration change request & receipt
- Review of requested change
- Approval/rejection of change
- Changes (in the case of approval)
- Confirmation of changes (in the case of approval)
- Suitability and acceptability of safety status
- Suitability and acceptability of system requirements status

Configuration information for status accounting

Configuration Status Accounting (CSA) is the process that helps track and monitor changes made to the system configuration. It provides a record of all changes made to the system and ensures that only authorised changes are made.

When the configuration information is documented, the following information should also be recorded:

- Configuration information (such as identification number, title, effective dates, revision status, change history, and its inclusion in any baseline)
- Status of release of latest configuration information
- Processing of changes.

The CSA information is used for updating the configuration management database and generating reports that summarise the changes made to the system configuration. This information can also be used for determining the impact of changes on other parts of the system and for identifying any unintended consequences of the changes.

< Status accounting reporting >
If there is any revision of configuration after the first baseline has been established, the configuration managers of the subcontractors should record the changes and report them periodically to the configuration manager at the system level. All configuration status should be synchronised between the subcontractor's configuration database and the configuration management database at the system level regularly.

Configuration audit

Configuration audit process

To verify the soundness of configuration management, a series of the configuration audits should be performed during the configuration management lifecycle. Configuration audits consist of three phases and their activities as follows:

- Audit preparation phase: Defining objectives of audit, assigning the audit team, preparing the audit agenda, and issuing the audit agenda and schedule.
- Audit execution phase: Holding the kick-off meeting for the audit, performing the audit, holding close-out meeting, and confirming findings and corrective action.
- Audit corrective action phase: Issuing the audit report and corrective action, performing corrective actions, and issuing the audit closure report.

The importance of configuration audits lies in their ability to verify that configuration items and their associated documentation are accurate, complete, and consistent with the specified requirements. They ensure that changes have been properly implemented, authorised, and recorded, helping to maintain system integrity and prevent issues caused by incorrect or unauthorised configurations. This process reduces risks and enhances the reliability and quality of the product or system.

Configuration audit

The configuration audits will be undertaken for both functional and physical configuration audit as follows:

- Physical Configuration Audit (PCA): To verify the established physical configuration items, ensuring that the identified initial system configuration is in alignment with the "As Built" configuration and meets the technical requirements.
- Functional Configuration Audit (FCA): To verify the achievement of the performance specified in the functional and/or allocated configuration identification.

However, FCA is seldom conducted because the verification and validation activities of other SE part (e.g., RAM, safety, requirements, EMC, etc.) typically cover it.

During the configuration audit, the following materials should be examined:

- System configuration worksheets
- Documents master lists
- Drawings master lists
- Configuration management records and their interrelations
- Other supporting evidence for configuration management, if available

EMC (ELECTRO-MAGNETIC COMPATIBILITY)

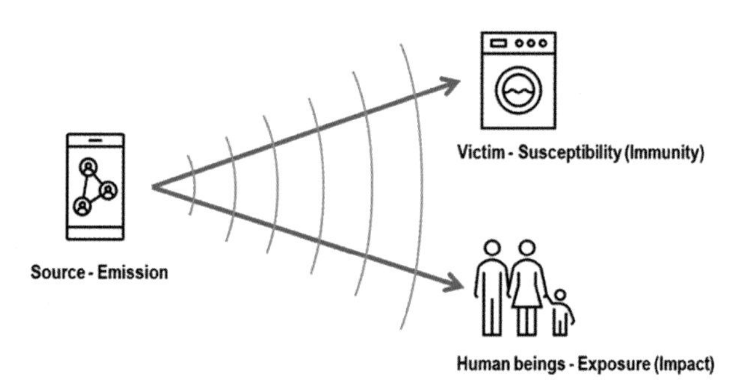

Overview of EMC

The electric field and magnetic field are closely related and interact through electromagnetic phenomena. When an electric charge moves, it creates a magnetic field around it, and when a magnetic field changes, it induces an electric field. This relationship is described by Maxwell's equations and is fundamental to electromagnetism. Together, electric and magnetic fields propagate as electromagnetic waves (such as light), where the changing electric field generates a magnetic field, and the changing magnetic field, in turn, generates an electric field, allowing the wave to travel through space. Consequently, any device or system powered by electricity is inherently subject to EMC (Electro-Magnetic Compatibility) issues.

EMC refers to the ability of electronic devices or systems to function properly within their electromagnetic environment without causing or experiencing interference. It ensures that devices do not emit excessive electromagnetic noise (EMI, Electro-Magnetic Interference) that could disrupt other equipment, while also being resistant to external electromagnetic disturbances. This ensures reliable and harmonious operation in shared environments.

Characteristics of electricity

To understand EMC management, it is essential to first grasp the fundamental characteristics of electromagnetism. Here are some characteristics that we need to know:

< Electromagnet >
When an electric current flows through a coil of wire, it generates a magnetic field like that of a magnet. Electromagnets are widely used in various applications, such as electric motors, transformers, and magnetic cranes, due to their controllable magnetic properties.

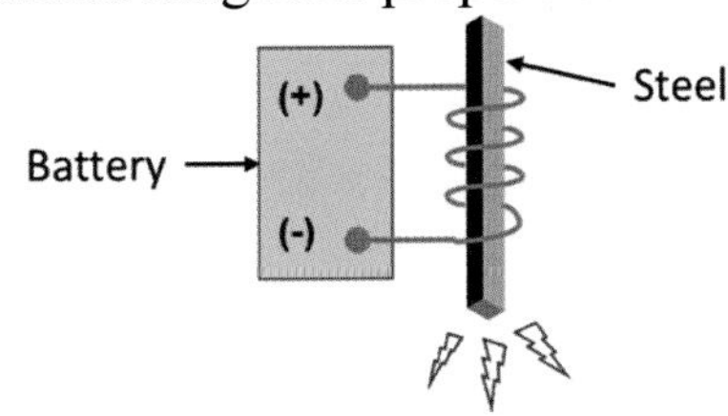

< Electromagnetic induction >
A changing magnetic field induces an electric current in a conductor. This occurs when either the conductor moves through a magnetic field or the magnetic field around a stationary conductor change. This principle underlines many technologies, such as electric generators and transformers, where mechanical energy is converted into electrical energy or vice versa.

< Transferring electrical energy >
A transformer consists of two coils of wire, known as the primary and secondary windings, wound around a core. When an alternating current (AC) flows through the primary coil, it generates a changing magnetic field that induces a voltage in the secondary coil.

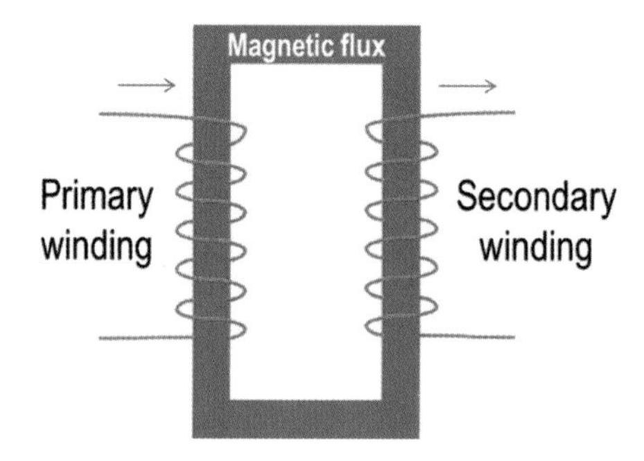

< Frequency spectrum >
Electromagnetic energy exists over a wide range of frequencies, from low-frequency power lines to high-frequency radio waves. Devices need to be designed to operate within specified frequency bands to avoid interference.

EMC issues

EMC issues can be classified into three categories: emission, susceptibility, and human impacts, as follows:

- Emission refers to the generation of electromagnetic energy. The main concerns regarding emission are unwanted emissions that may cause interference with other equipment, the environment, or human beings. It primarily focuses on sources, and EMC studies the countermeasures that can be implemented to reduce unwanted emissions.

- Susceptibility refers to the tendency of equipment to malfunction or fail in the presence of unwanted emissions, also known as Radio Frequency Interference (RFI). Immunity is the opposite of susceptibility, describing the ability of equipment to function correctly in the presence of RFI. It mainly concerns the victims of interference. EMC studies the design techniques that can enhance the immunity of equipment.
- The risks to human beings caused by time-varying electromagnetic fields (EMF), static magnetic fields, and electric shock should be assessed. If not properly managed, humans may be exposed to unduly high levels of EMF, which could impact on their health.

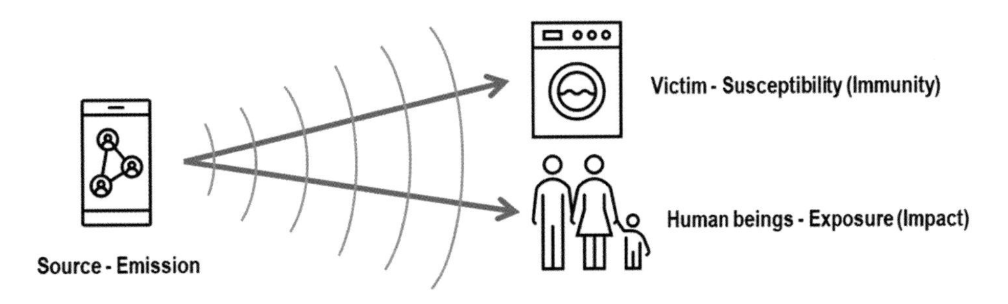

Here are the target subsystems relating to EMC issues in railway projects:

- Rolling stock and its interface with the catenary or third rail must be designed considering both emission and susceptibility in relation to other subsystems.
- Signalling systems consist of equipment distributed along the track, installed in the driver's cab of rolling stock, and at the operation control centre.
- Communication systems comprise equipment distributed along the track and installed wherever any type of communication modes is needed, such as the operation control centre, radio trackside, etc.
- Power supply and distribution systems are made up of transformers and rectifiers dispatched along the line. The harmonics generated by the systems may be critical for signalling system.

Additionally, the external (or environmental) E&M systems, which are expected to have a potential impact on performance of the railway systems, should be regarded as external constraints during the design phase.

EMC-related standards

The standards related to EMC are as follows:

- EN 50121-1: Railway applications. Electromagnetic compatibility. General
- EN 50121-2: Railway applications. Electromagnetic compatibility. Emission of the whole railway system to the outside world
- EN 50121-3-1: Railway applications. Electromagnetic compatibility. Rolling stock. Train and complete vehicle
- EN 50121-3-2: Railway applications. Electromagnetic compatibility. Rolling stock. Apparatus
- EN 50121-4: Railway applications - Electromagnetic compatibility, Emission and immunity of the signalling and telecommunications apparatus
- EN 50121-5: Railway applications. Electromagnetic compatibility. Emission and immunity of fixed power supply installations and apparatus
- EN 50122: Railway applications – fixed installations Part 1 Protective provisions relating to electrical safety and earthing
- EN 50155: Railway applications. Electronic equipment used on rolling stock
- EN 61000-6-1: Electromagnetic compatibility (EMC). Generic standards. Immunity for residential, commercial and light-industrial environments
- EN 61000-6-2: Electromagnetic compatibility (EMC). Generic standards. Immunity for industrial environments
- EN 61000-6-3: Electromagnetic compatibility (EMC). Generic standards. Emission standard for residential, commercial and light-industrial environments
- EN 61000-6-4: Electromagnetic compatibility (EMC). Generic standards. Emission standard for industrial environments

EMC Management

The purpose of EMC management activities is to control all the EMC aspects throughout the lifecycle of the project to deliver compliant, high-performance, and safe systems. The EMC Management Plan should define the following:

- Overall EMC management and organizations
- Overview of the EMC design and control methodologies
- Standards to be applied
- EMC and safety assurance
- Change control process
- EMC interfaces management
- Operational constraints
- Activity plans
- EMC testing and commissioning

Target systems

In the case of DC metro with third rail, the following list includes the subsystems to be considered during the design phase:

- Power supply & distribution system: DC substations (including transformers, rectifiers, DC circuit breakers, switchboards, LV panels, and feed cables), connections to SCADA, third rail conductors, cables, battery, and UPS systems
- Signalling: On-board ATP equipment in driver's cab, ATP equipment at the wayside, ATS equipment at the signalling control centre, communication devices to provide communication between on-board/wayside/central control, track circuit, UPS for signalling
- Communications & Automatic Fare Collecting (AFC) system: fibre optic communication backbone with transmission nodes (for voice, data, and video), telephone systems for administration, telephone systems for emergencies (IP-PBX), CCTV, trunk radio systems, master clock systems, public address systems, train radio systems, Maintenance Management System (MMS), automatic ticket vending machines, central computer system for AFC, workstations in CCR, and SCADA

- Rolling stock: VVVF inverters, traction motors per car, auxiliary inverters (SIV), battery chargers, electro-pneumatic braking, door systems, lighting, HVAC, PA/PIS
- Rolling stock maintenance equipment: machines in heavy maintenance and light maintenance shops, and train washing plants

The following steps are the commonly used EMC management process:

- Perform assessment to characterise the site environment.
- Establish EMC criteria to apply in design.
- Identify systems that might have EMC impact on other systems or be affected by others.
- Check whether the EMC-related certificates of each system are appropriate to EMC requirements.
- Conduct validation activities on the systems to demonstrate that they meet the EMC requirements.

Interface management for EMC control

Interface management for EMC control involves managing the interfaces between electronic devices and systems to ensure that they operate correctly in their intended electromagnetic environment. The following activities are key considerations subsystem suppliers when managing interfaces for EMC:

- Identification of the EMC interfaces: The first step is to identify the interfaces between different electronic devices and systems. This includes the interfaces between the equipment and the external environment, as well as between different internal components.
- Determination of the electromagnetic environment: The next step is to determine the electromagnetic environment in which the devices will operate. This includes identifying potential sources of electromagnetic interference (EMI) and determining the levels of electromagnetic waves that the devices are allowed to be exposed to.
- Identification of interface requirements: Once the interfaces and electromagnetic environment have been identified, the interface requirements should be defined. This includes determining the acceptable levels of EMI and ensuring that the interfaces are designed to meet these requirements.
- Designing and testing the interfaces: The next step is to design the interfaces and test them to ensure they meet the interface requirements. This may involve using simulation tools to analyse the electromagnetic behaviour of the interfaces, as well as conducting physical tests to verify their performance.
- Implementation and maintenance of the interfaces: Finally, it is necessary to implement and maintain the interfaces to ensure they continue to meet the interface requirements over time. This may involve monitoring the performance of the interfaces, conducting periodic tests to verify their performance, and making modifications as necessary.

Site assessment and measurement

The purpose of the site assessment is to measure the level of the RFI (Radio Frequency Interference) around the railway line and predict its impact once the railway line is installed. The assessment provides appropriate guidelines for designing the railway systems.

Environment measurement should be carried out in accordance with EN 50121-2. The measurement instruments should include antennas covering the frequency spectrum from 9 kHz to 3.6 GHz, RFI receivers, and EMF meters. It should be verified that all the test instruments have a valid calibration certificate.

Figure 82 - An example of measurement result

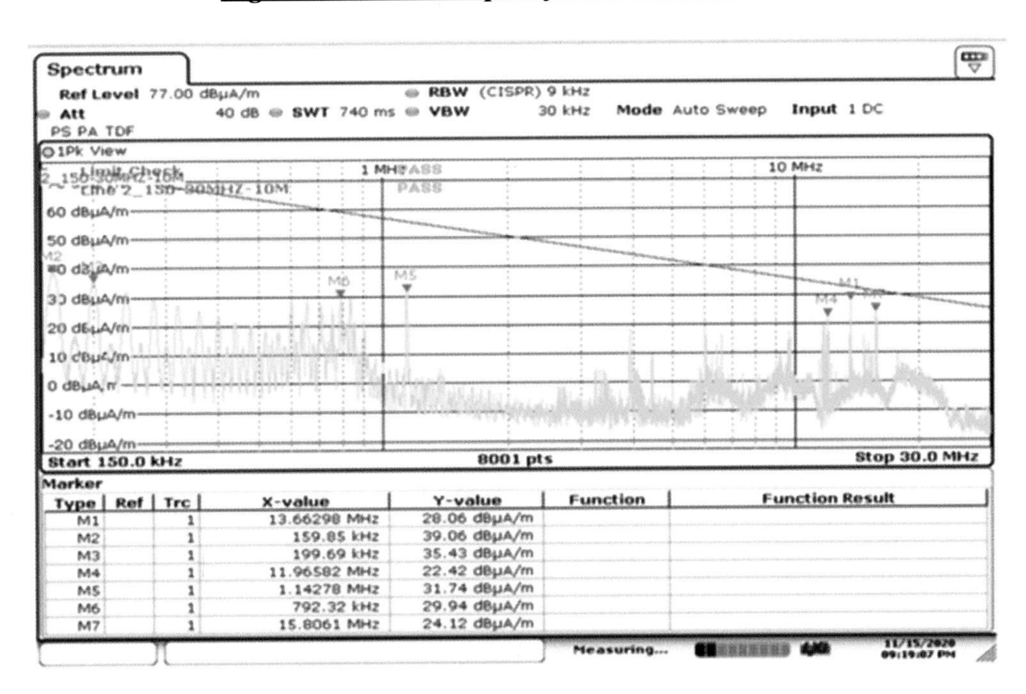

EMC analysis

< Inter-systems >

Inter-system analysis involves analysing EMC issues between the systems of the subcontractors. Since each subcontractor is focused solely on their own systems, an EMC manager is required to manage issues between them.

EMC analysis of inter-systems demonstrates each equipment's functionality in terms of the following aspects:

- A component within railway systems should not generate electromagnetic waves or currents above the acceptable levels specified in the standards, which could impact on neighbouring components within the railway systems.
- A component within railway systems should operate properly despite the electromagnetic waves or currents generated by other components in the railway systems, if they remain within the acceptable level defined in standards.

< Extra-systems >

EMC analysis of extra-systems demonstrates the functioning of the railway systems with respect to their impact on equipment outside the railway system in terms of the following aspects:

- The railway system should not induce significant electromagnetic disturbances to equipment outside the railway system.
- The railways system should operate correctly despite the electromagnetic waves and currents generated by equipment outside the railway system.

There might be claims from individuals, communities, or organisations outside the project that their sensitive equipment is malfunctioning due to the newly opened railway (your project). Whether these claims are valid or not, they can cause issues for rail operators. Therefore, the EMC manager should conduct an EMC site assessment at the start of the project and during the trial operation phase. Both assessment records and analysis will be used for demonstrating that the new railway does not impact neighbouring facilities or equipment in terms of electromagnetic wave.

Earthing, Bonding, and Cabling

Definitions

The terms ‘earthing’ and ‘grounding’ are often used interchangeably, but they have slight differences depending on the region:

- The word “earthing” is commonly used in the UK and refers to connecting the electrical system to the earth (ground) for safety, ensuring that any fault current is safely discharged into the ground to prevent electric shock.
- “Grounding” is the term used in the US and refers to the same concept but can also mean connecting parts of an electrical circuit to a common reference point (often the ground) for stability and protection.

Bonding in electrical systems refers to the process of connecting all metallic parts that are not intended to carry current (such as pipes, metal enclosures, or structural steel) to the same electrical potential by using conductors. This ensures that there are no voltage differences between these parts, thereby reducing the risk of electric shock in the event of a fault. Bonding works in conjunction with earthing to enhance electrical safety by minimising the risk of electrical hazards.

Cabling refers to the installation and organisation of cables to connect devices or systems for the transmission of electrical power, signals, or data. This includes various types of cables, such as electrical cables, fibre optics, or network cables. Proper cabling ensures reliable and efficient communication and power distribution while minimising interference, signal loss, and potential hazards.

Classification of cables and cabling guidelines

According to IEC standards, cables can be classified as follows:

- Class 1 (Very sensitive signals): High-rate digital and analogue communications, such as ethernet, video, RFI receiver cables from antennas.
- Class 2 (Slightly sensitive signals): Ordinary analogue 4 to 20 mA and 0 to 10 V signals (less than 1 MHz) and low-rate digital bus communications.
- Class 3 (Slightly interfering signals): Control circuits and induction motors.

- Class 4 (Strong interfering signals): Cables for the LV supply and control cables to unsuppressed inductive loads.
- Class 5 and 6: Power distribution cables.

Here are general design considerations for cabling:

- It is essential to identify the class of cable.
- Power cable should not share ducts with other cables.
- A parallel earthing conductor is recommended.
- All shielded cables should be earthed.
- Each cable should be installed separately with appropriate spacing.
- Power circuits and sensitive circuits should cross at right angles.

Here are general design considerations for cable separation:

- It is important to provide shielding or physical separation from magnetic fields generated by power cables, especially at power frequencies and their harmonics.
- Signalling or communication cables should be kept separated from high-current power cables, and their conduits, as sensitive cables need to be shielded to protect them from interference.
- Sensitive equipment should be separated from sources of interference to ensure safety.
- If it is not possible to install a barrier between sensitive equipment and sources of interference, the distance between them should be maximised.
- The spacing between two adjacent classes of cables in air should be at least 10 times the diameter of the larger bundle.

Guidelines for earthing & bonding and shielding

Here are the general design considerations for earthing and bonding:

- Lightning rods and sensitive devices should be installed away from each other.
- All exposed metallic parts should be connected to the earthing network to ensure safety and EMC performance.
- The materials used for the earthing system and bonding straps should be same, as dissimilar materials can lead to electrochemical effects.

- Suitable conductors for bonding straps include metal strips, metal mesh straps, or round cables. For high-frequency systems, metal strips or braided straps are preferred, with a typical length-to-width ratio of less than five.
- Grounding connections require clean metal surfaces, so paint or non-conductive layers should be removed from contact areas.

Here are the general design considerations for shielding:

- Shielding is necessary when adequate separation between sources of EMI and sensitive cables is not possible.
- Steel is the preferred shielding material for conduits due to its high permeability and skin effect, which helps dampen currents.
- Deep conduits are preferred because they have lower mutual inductance.

Human Exposure

The analysis of human exposure aims to assess the risk posed by electromagnetic fields to human beings.

Requirements for human exposure

< Electromagnetic fields >
ICNIRP (International Commission on Non-Ionizing Radiation Protection) guidelines set the exposure limit for 50 Hz (power frequency) at 10 mA/m^2 for occupational settings and 2 mA/m^2 for the public, based on WHO studies. These levels, known as basic restrictions, are determined from observation that minimal biological effects occur at body current densities from 10 mA/m^2 to 1 mA/m^2, and a density of 2 mA/m^2 considered safe for human health.
The basic restriction on body current densities is challenging to measure directly in humans. Therefore, ICNIRP provides reference levels for time-varying electric and magnetic fields, which can be measured using portable equipment. These reference levels assist in assessing compliance with basic restrictions.

< Static magnetic fields >
According to ICNIRP guidelines, transient exposure to static magnetic fields up to 2 Tesla does not show harmful effects on major developmental, behavioural, or physiological parameters in higher organisms. The ICNIRP limits for static magnetic field exposure are detailed in Figure 83, and measurements are conducted according to the requirements outlined in the EN 50500 standards.

Figure 83 - Magnetic flux density

	Exposure type	Magnetic flux density
Occupational	Exposure of hand	2T
	Exposure of limbs	8T
Public	Exposure of any part of body	400 mT

< Medical implantation and wearable devices >
Human exposure analysis must consider individuals with implanted medical devices. Tests on heart pacemakers have shown that static magnetic fields can affect the pacing rate, but exposure below 500 μT is unlikely to cause issues. Additionally, interference with pacemakers from magnetic fields is unlikely below 100 μT or 200 μT.
EN 50121-5 specifies an immunity level for fixed power equipment of 300 A/m (377 μT) in the railway environment. This standard indicates that such field strengths may be expected in equipment rooms.
High harmonic content in DC traction currents during acceleration or regenerative braking can interfere with hearing aids by causing high magnetic field emissions. This interference degrades the signal-to-noise (S/N) ratio of hearing devices. According to EN 60118-4, an acceptable S/N ratio is 32 dB, though a 22 dB ratio is also considered acceptable for short periods if the magnetic noise is not noticeably tonal.

Human exposure assessment in railway sites

< Accessible areas for workers inside the rolling stock >
Measurements should be taken near train emission sources where workers typically operate, including the driver's seat. They should be conducted at heights of 0.9 m and 1.5 m above the floor, and at least 0.3 m from walls or the minimum distance where workers are present.

< Public areas inside the rolling stock >
Measurements should be taken as close as possible to train emission sources accessible to the public. In public areas, measurements should be conducted at heights of 0.3 m, 0.9 m, and 1.5 m above the floor, and at least 0.3 m from walls or the minimum distance where the public is present.

< Fixed power supply >
To demonstrate compliance with the existing railway infrastructure, measurements must be taken at heights of 0.3 m, 0.9 m, and 1.5 m in public areas, and at 0.9 m and 1.5 m in worker areas. Measurements should be conducted at least 0.3 m from walls or fences, or the minimum distance where people are present.

< Platform >
On the platform, measurements should be taken at heights of 0.9 m and 1.5 m above the platform level, and at a horizontal distance of 0.3 m from the platform edge.

SOFTWARE ASSURANCE

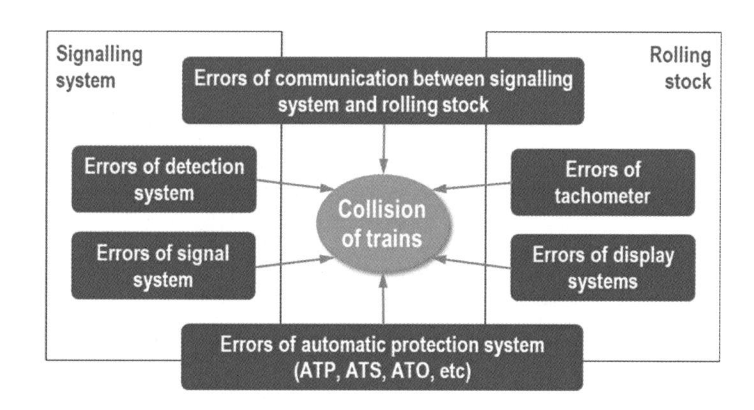

Overview of software assurance

The purpose of software assurance is to ensure that software systems and applications are designed, developed, and maintained to meet quality standards, security requirements, and reliability expectations. It encompasses a comprehensive set of processes and methodologies to minimise risks, detect vulnerabilities, and prevent defects, ensuring that the software functions as intended, remains secure against threats, and performs reliably under various conditions.
While hardware can face various types of risks except for intrinsic cybersecurity failures, software can experience both systematic failures and cybersecurity vulnerabilities, as shown in Figure 84. Thus, software development requires different approaches from hardware development.

Figure 84 - Risk types of HW and SW

Type of risks	Hardware	Software
Systematic failure	v	v
Random failure	v	
Physical security failure	v	
Cyber security failure		v

Software assurance in Systems Engineering

The software assurance activities focus on systematically managing and verifying software development at each lifecycle phase as part of Systems Engineering efforts, ensuring that all software is developed in accordance with the standards and quality & safety requirements predefined in the Systems Engineering Management Plan (SEMP) or System Assurance Plan (SAP) of a project. Therefore, software management methodologies should align with the systems engineering lifecycle methodology described in the SEPM or SAP.

Railway subsystems or components can be divided into two main categories: hardware and software. However, systems engineering activities are often approached primarily from a hardware perspective. With the digitisation of many railway systems, software assurance must also be considered for each system. Figure 85 illustrates the scope of systems engineering activities on railway hardware and software. As shown, software is involved in all SE activities except RAM and EMC. Consequently, software assurance should be regarded as a critical aspect of SE activities.

Figure 85 – Coverage of systems engineering activties

SE Area	Hardware	Software
Requirements	v	v
V&V	**v**	**v**
Configuration	v	v
Interface	**v**	**v**
RAM	v	
Safety	**v**	**v**
Human Factors	v	v
EMC	**v**	

Software assurance process

As a software assurance manager at the top level in the organisational structure, the primary tasks include reviewing software deliverables produced by suppliers and auditing to verify conformity with software development process defined by ISO 90003 and EN 50128.

Typical key activities of suppliers involved in software development processes include:

- Requirements gathering and analysis: Identifying software development requirements and ensuring these requirements are feasible, testable, and aligned with the overall project goals.
- Software design and development: Designing the software architecture and developing the code to meet the requirements, typically using software development methodologies, such as Agile or Waterfall to organise and manage the work.
- Testing: Conducting thorough testing and validation to identify and resolve defects before deployment.
- Deployment and maintenance: Deploying the software to the production environment and providing ongoing support and maintenance to address issues and enhance the software over time.
- Risk management: Identifying and mitigating risks associated with the software development and deployment process, such as quality concerns.

However, much of the software used in railway systems is COTS (Commercial Off-The-Shelf). Proven software is preferred, and a short development period is required. Even when COTS software is not used, only software proved in other projects is typically implemented, with customisation achieved through parameters adjustments. As a result, the suppliers activities mentioned above for software development are rarely observed in rail projects.

Software quality assurance

In railway projects, suppliers developing software are required to be accredited to ISO 9001, and to follow the software engineering practices outlined in ISO 90003, which include:

- Defining software quality goal and metrics.
- Measuring and monitoring software quality.
- Managing software configuration and changes.
- Applying software development methodologies for design and testing.
- Specifying requirements for designing and testing software.
- Conducting software verification and validation.

Software Assurance

Figure 86 outlines the software assurance activities performed by SE team, subcontractors, and sub-subcontractors.

Figure 86 – Software assurance activities

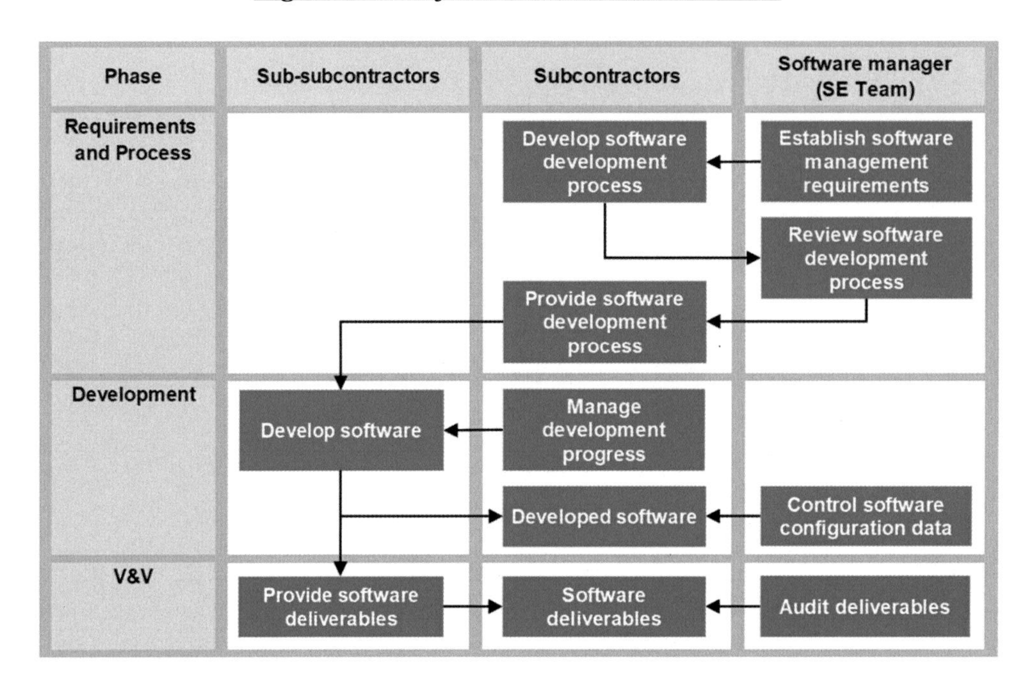

At system level:

- Establish software management strategies and requirements
- Provide software management requirements to suppliers for system-level software assurance.
- Review suppliers' software development process and progress.
- Manage and control suppliers' software configuration items, including a configuration data or filed parameter data.
- Audit suppliers' deliverables to verify conformity with standards and defined software development processes.

At subsystem level:

- Manage and control the software development progress for subsystem components and equipment.
- Monitor software quality (attributes) trend of subsystem components or equipment.
- Compile software deliverables for components or equipment at the subsystem level, if necessary.

At component level:

- Develop software in accordance with the requirements stated in the software management plan and the software requirements specification, applying best software engineering practice as per standards.
- Provide software deliverables.

Software quality assurance and configuration management

The software assurance is a planned and systematic set of activities that ensure software lifecycle processes and products conform to requirements, standards and procedures. Software quality assurance, on the other hand, is a function of software quality that assures the standards, processes, and procedures are appropriate for the project and correctly implemented. The objective of software quality assurance is to identify, monitor, and control the technical and managerial activities necessary to ensure that the software meets the required quality standards.
To achieve the objectives of software quality assurance, suppliers should develop planning documents at the outset of the project, including software quality assurance plan, software configuration management plan, and software verification and validation plan. These documents should be updated throughout all lifecycle phases as live documents. This ensures that a common software quality assurance framework is established.

< Software configuration management >
The software configuration management ensures the consistency of developed software deliverables as they change. Suppliers should prepare a software configuration management plan for their software based on their own quality management system and standards such as ISO 10007 and IEEE 828. These standards primarily cover the following:

- Identification of software configuration items and the baselines.
- Change control.
- Configuration status accounting.
- Configuration audits and reviews.

Software configuration management should control all software configuration items, particularly suppliers' software-related documents, including source code, executable files, configuration data, and software release information. These should be inspected and audited by the system software manager.

Software development

< Software design, development and testing >
The main software development activities can be summarised in the following phases: software design, development, and testing. Software design encompasses software requirements, software architecture, software interfaces, and software component design.
Considering the features of the software language and target hardware/system, appropriate software design methodologies should be chosen based on the desired quality attributes.

< Software development tools >
Software development tools should be selected as an integral part of the software assurance activities to prevent potential tool failures from adversely affecting the output of the integrated toolset.
All tools to be used for software development should be fit for purpose, with their integrity assured and demonstrated to be compatible with the application's requirements. All tools should be calibrated, if applicable, or have a specification or manual that clearly defines the tool's behaviour, along with any instruction or constraints on its use.
Adequate evidence for the tool's use should also be provided, which may be based on:

- A proven history of successful use in similar environments and for similar applications.
- Redundant code that allows for the detection and control of failures resulting from faults introduced by the tool.
- Application constraints or additional measures derived from risk analysis of the process and procedures, including the tools.
- Other appropriate methods for preventing or handling failures introduced by tools.

< Software development techniques >
For software development, the following techniques can be used to ensure that the software quality is achieved:

- Software requirement specification: Structured methodology and Modelling.

- Software architecture, design, and development: Defensive programming, fault detection & diagnosis, fully defined interface, structured methodology, modular approach, and design & coding standards
- Verification, testing, integration, and overall software testing: Static analysis, dynamic analysis & testing, traceability, test coverage for code, functional & black-box testing, performance testing, and interface testing

Software integration and V&V

During the software testing phase, the software should be integrated, considering the interface and interaction between software components (i.e., software modules) and the target hardware. The integration of software components involves progressively combining them so that their interfaces and the assembled software can be adequately verified before system integration and testing.

Software integration must be performed, considering the interaction between software components and the target hardware, after the completion of software unit testing.

< Software verification and validation >

Through the software V&V processes, it is determined whether the software satisfies the intended use, user needs, and software requirements. This determination involves assessment, analysis, evaluation, review, inspection, and testing of software products and processes. Software V&V may be performed in parallel with the software development.

To achieve this V&V objective, the following should be checked:

- Evaluation of software design, from software requirements specification to software component design specifications, including corresponding test specifications.
- Correctness and consistency of the input to that phase.
- Traceability analysis.
- Conformity to standards and planning documents.

The review and audit will be the main tasks for software management at system level, assessing the deliverables submitted by suppliers as part of software V&V activities.

Software Assurance Plan

The purpose of the Software Assurance Management Plan is to establish, document, and maintain a consistent approach for software development management, as carried out by a contractor for a project. It also provides all subsystem suppliers and their sub-suppliers with an overview of the contractor's approach to managing software quality in accordance with the design & engineering management plan and the system assurance plan.

Scope of work for railway projects

Subsystems that adopt software are as follows:

- Signalling system: electronic interlocking equipment, local control console, AF (Audio Frequency) track circuit, OCC (Operating Control Centre)
- Communications and AFC (Automatic Fare Collection) system: transmission system, radio system, public address (PA) system, CCTV system, electronic clock system
- Power supply and distribution system: SCADA (Supervisory Control And Data Acquisition)
- Rolling stock: doors, HVAC (Heating, Ventilation, and Air Conditioning), friction brake system, traction system, train communication system & passenger information, train control & monitoring system, vehicle on-board controller (train control management system), and other man-machine interfaces.

Here are some references that can be used for developing a software assurance plan:

- ISO 9001: Quality management systems – Requirements
- ISO 10007: Guidelines for configuration management
- ISO/IEC 90003: Software Engineering – Guideline for the application of ISO 9001 to computer software
- ISO/IEC/IEEE 15288: Systems and software engineering – System life cycle processes
- BS EN 50128: Railway Applications – Communication, signalling and processing systems – Software for railway control and protection systems
- IEEE Std828: Standard for configuration management in systems and software engineering

Roles and responsibilities

A system software manager at the top level in the organisational structure is responsible for overseeing software management activities throughout the project lifecycle, as well as integrating subsystems provided by subcontractors.

The system software manager should:

- Manage software requirements at the system level
- Set up the software management strategy & requirements and prepare the software assurance management plan
- Report software quality, safety issues, and software management status to the system integrator
- Review and monitor the supplier's software development status
- Review key software deliverables prepared by suppliers
- Prepare the software assurance reports for software design, software validation and final stage

Subsystem software managers are responsible for software management activities at the subsystem level throughout the project lifecycle. They should:

- Communicate and interface with the system software manager
- Provide software requirements to their sub-suppliers, if required
- Prepare software deliverables
- Report software quality and safety issues to the system software manager

Software safety classification

Software should be managed according to its safety classification. Software safety can be classified as follows:

- Non-Safety: Software failures that do not lead to any dangerous situation, and the associated risks are negligible.
- Safety Class 1: Software failures that may lead to a dangerous situation, and the risks must be reduced to an acceptable level.
- Safety Class 2: Software failures that will lead to a dangerous situation, and the risks must be eliminated.

The safety classification of rolling stock systems is as follows:

- Safety Class 1: Doors & gangways, and friction brake system
- Safety Class 2: Vehicle on-board control system (train control management system)

They safety classification of signalling systems is as follows:

- Safety Class 1: Local control console, OCC (Operating Control Centre)
- Safety Class 2: Electronic interlocking equipment, AF (Audio Frequency) track circuit

Software should also be developed to meet system and safety requirements in accordance with the safety standard EN 50128. In addition to the basic software quality features, safety-related software should be developed with particular attention to the following:

- Software safety requirements and their test specifications.
- Organisational independence between design & testing, and software V&V.
- Techniques and measures adopted to meet the safety requirements.

Software development

Software design

Software development plans should be developed by suppliers at the beginning of software development to address the overall requirements, which will be used for structuring the Software Quality Assurance Plan (SQAP), Software Configuration Management Plan (SCMP), and Software Verification and Validation Plan (SVVP). The software should be designed in phases, applying good software engineering practices as specified in the ISO 90003 and EN 50128 standards.
The activities in each lifecycle phases are:

- Software requirements phase: Describe a complete set of requirements for the software that meet all safety requirements and provide a comprehensive set of documents for each subsequent phase.
- Software architecture and design phase: Develop a software architecture and design that meet the software requirements; identify and evaluate the significance of hardware/software interactions.
- Software component design phase: Develop a software component design that meets the requirements of the software design specification; develop a software component test specification that meets the requirements of the software component design specification.
- Software component implementation phase: Develop software that is analysable, testable, verifiable and maintainable. The software should be designed to meet these objectives by applying appropriate methodologies in this phase.

Software testing and validation

Software testing should be performed in phases, based on the test specifications prepared at each software design phase, such as the software component test specification, software integration test specification, and software requirements test specification.
The objectives of each software development lifecycle phase are as follows:

- Software component testing phase: Test software components to verify that each component meets the requirements and functions as intended.
- Software integration phase: Perform software-hardware integration and demonstrate that the software and hardware interact correctly to perform their intended functions.

- Software validation phase: Analyse and test the integrated software and hardware to ensure compliance with the software requirements specification, with particular emphasis on the functional and safety aspects according to the software safety integrity level and verify that it is fit for its intended application.

A software validation report should be produced to provide a summary of the tests results and confirm whether the entire software on its target equipment fulfils the requirements set out in the software requirement specification (SWRS). This should include an evaluation of the test coverage on the requirements of the SWRS, an assessment of other verification activities in accordance with the software verification and validation plan, and confirmation that appropriate techniques and measures have been applied.

Software maintenance and optimisation

Software maintenance and optimisation may be considered part of software validation in accordance with the software development process defined by suppliers. Therefore, software validation may be finalised during this phase.

This phase primarily focuses on software changes and optimisation after installation on site. All software should be adequately tested before site installation at the supplier's premises (e.g., factory acceptance test) and finally tested on site (e.g., site acceptance test). If software modification or optimisation occur during this phase, they should be changed or updated in accordance with pre-defined software maintenance plans or procedures, as well as the software configuration management plan, to maintain the required software quality and safety as originally designed.

Cyber security

Why cyber security

Many systems in railways have been digitised, creating extensive networks interconnected through communication systems. While digitisation has improved the efficient of railway systems, it has also heightened the risks of cyber-attacks.
From the perspective of attackers, such as terrorists, a cyber-attack on rail network systems is a low-cost method for carrying out attacks. There is also a reduced risk of being caught on-site compared to conducting a physical attack, making cyber-attack a more likely choice over traditional terrorism.

On the other hand, it is essential first to define what cyber-attack is. As we have identified the types of risks, there are two types of security failures: physical security failures and cybersecurity failures. However, distinguishing between these two types can be challenging when a security failure occurs. Therefore, in this chapter, a cyber-attack is defined as an attack where trespassers primarily exploit railway communication systems or digitalised systems for offensive purposes.

Key aspects

Key aspects to manage for ensuring the safety and security of software in railway systems include:

- Network security: Protect the rail network from cyber threats by implementing robust network defences such as firewalls, intrusion detection systems (IDS), and intrusion prevention systems (IPS). Secure communication channels and network segmentation are essential prevent unauthorized access.
- Critical infrastructure protection: Secure critical components of rail infrastructure, including signalling systems, control centres, and communication networks. These systems are crucial for safe and efficient rail operations and must be protected from cyberattacks.
- Access control: Enforce strict access controls to ensure that only authorised personnel can access sensitive systems and data. This involves managing user permissions, implementing strong authentication methods, and regularly reviewing access rights.

- Incident response: Establish and maintain a comprehensive incident response plan to address cybersecurity incidents promptly and effectively. This plan should include procedures for threat detection, analysis, and mitigation, along with communication protocols for notifying stakeholders.
- Vulnerability management: Perform regularly vulnerability assessments and penetration testing to identify and resolve security weaknesses in rail systems. Timely patching and updates are critical to protect against known vulnerabilities.
- Data protection: Maintain the confidentiality, integrity, and availability of data used in rail operations. This includes securing sensitive data from unauthorized access and ensuring data backup and recoverability in case of a security breach.
- Compliance and standards: Follow relevant cybersecurity standards and regulations tailored to the rail industry. This includes adhering to industry guidelines, national regulations, and international standards to maintain compliance and best practices.
- Employee training: Conduct regular cybersecurity training for employees to enhance awareness of potential threats and reinforce best security practices, as human error remains a critical risk factor.
- Continuous monitoring: Use continuous system monitoring and logging to detect and respond to suspicious activities in real-time, enabling early threat identification and timely mitigate measures.

Targets and vulnerabilities in railway systems

The scope of cybersecurity in railways encompasses the protection of the infrastructure, systems, and data from cyber threats. This includes securing communication networks, signalling systems, operational control centres, onboard systems, and passenger data against hacking, malware, data breaches, and other cyberattacks that could disrupt operations or compromise safety.

When addressing cybersecurity issues in the field, it becomes evident that the scope of cybersecurity is extensive, making it challenging to define the appropriate level of depth required.
The author provides an example of a straightforward scenario to offer insight:
One of the objectives for cyber attackers may be to cause serious accidents within the railway systems, resulting in casualties and damage to infrastructure. Their aim extends beyond merely disrupting rail operations and causing inconvenience. From the perspective of cyber attackers, a train collision could be seen as a prime method to create significant catastrophic impact in railway systems.

Train collisions occur when the safe distance between a preceding train and a following train is not maintained. The distance between trains is calculated based on the speed of the following train, as indicated by the formula below:

$$\text{Distance} = \frac{\text{Velocity}^2}{2 \times \text{Deceleration rate} \times \text{Friction coefficient}}$$

If the speed of the following train is known, you can determine whether the distance between trains is sufficient. Conversely, if the distance is known, you can assess whether the speed of the following trains is appropriate. To maintain a safe distance, the following train's speed must be accurately measured, and to regulate its speed, the distance between the trains must be measured precisely.
Figure 87 shows the causal factors for train collisions. Detailing the weaknesses of the railway system in this book may inadvertently provide cyber attackers with insights into potential vulnerabilities, so a detailed description will be avoided. (Please note that the scenario largely depends on the type of signalling system in use.)

Figure 87 – Causal factors for train collisions

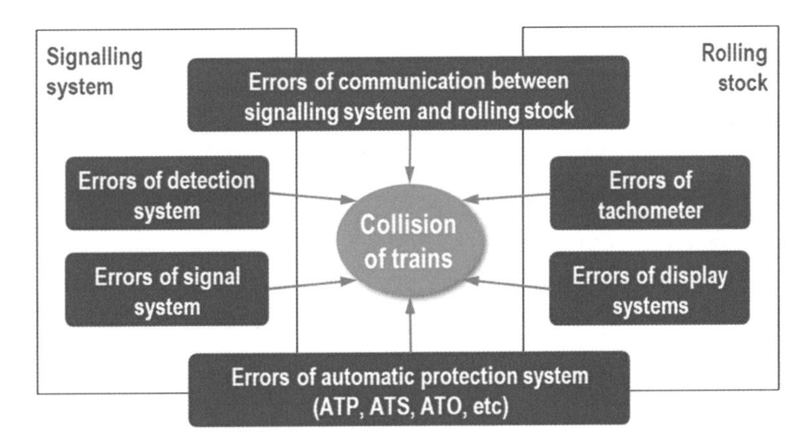

A tachometer (among various methods for measuring train speed) is a device used to measure a train's speed. If the device or speed display board in train cab fails, it could lead to significant issues due to human error. However, modern ATP or other automatic train protection systems can mitigate such failures.

Train detection systems, such as track circuit and axle counter, are used to measure the distance between trains. If these systems fail, the distance will be miscalculated, allowing two trains to enter the same block, which could lead to a collision. A block permits only one train at a time; otherwise, a collision could occur. Decades ago, many train drivers had no options but to rely solely on signal (traffic light) to control train speed, so signal system errors could cause significant issues.

< Enhancement of standards and requirements >

Cybersecurity requirements for the railway systems have yet to be fully standardised, and some projects demand excessive cybersecurity measures, significantly raising costs for facilities unintentionally. Although everyone desires perfect cybersecurity in railway systems, achieving this is costly; therefore, an optimised solution is essential.

One of the most important aspects of cybersecurity management is identifying system vulnerabilities. Although railway systems may appear broadly vulnerable, these vulnerabilities can be classified as either critical or non-critical. Vulnerabilities should be identified through scenarios-based cyberattacks assessments. Based on these assessments, railway standards or strategies relating to cybersecurity risks can be specifically developed.

To enhance standards and requirements suitable for cybersecurity in railway projects, the following steps are recommended:

- Refine the definition and scope of cybersecurity.
- Define the type of cybersecurity issues (key aspects) as introduced in the previous chapter.
- Develop scenarios relating to each issue, such as the example scenario introduced above.
- Identify and prioritise vulnerabilities (hazards) through assessment. Hazard analysis methodologies are covered in the chapter "Safety analysis".
- Develop criteria for cybersecurity management in railway projects.

Cybersecurity process

Analysis of various railway projects reveals that excessive cybersecurity requirements often stem from applying general cybersecurity requirements (e.g., those of IT systems) to railway systems. It is essential for clients and their advisors to assess the applicability of these requirements to railway projects. Unlike general IT systems, which connect to multiple networks and therefore require complex cybersecurity measures, railway systems operate on closed networks with operators typically using intranets. Thus, the cybersecurity requirements for railway systems do not need to align with those for IT systems.

Figure 88 outlines the process for optimising cybersecurity requirements specifically tailored to railway systems, as well as for managing cybersecurity risks within railway networks.

Figure 88 – Process of cybersecurity

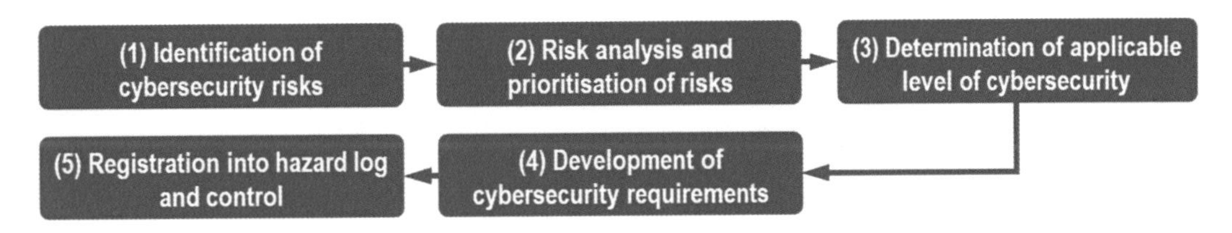

(1) Identify cybersecurity risks by conducting a thorough HAZID (HAZard IDentification) and PHA (Preliminary Hazard Analysis), like safety management. Skipping this process may lead to the unnecessary adoption of generic IT requirements, which many not be suitable for railway systems.

(2) Prioritise risks through risk analysis to focus the cybersecurity scope specifically on railway systems.

(3) Determine the appropriate cybersecurity level by applying the ALARP principle (As Low As Reasonably Practicable).

(4) Develop cybersecurity requirements based on the determined level of cybersecurity.

(5) Register all cybersecurity requirements into Hazard Log and manage them similarly to safety hazards.

NOISE AND VIBRATION

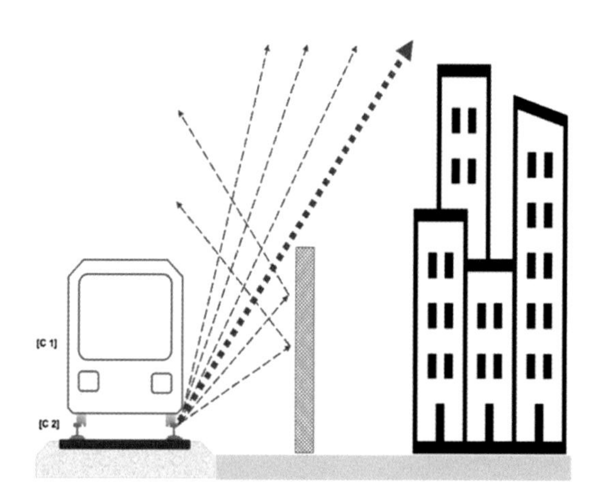

Overview of noise and vibration

Noise and vibration issues

With the recent rise in living standards, noise and vibration (N&V) requirements have become stricter, making it increasingly challenging to manage civil complaints related to N&V issues. Even when railway infrastructure builders meet the N&V regulations in each country, they may still need to install additional soundproof walls to address public complaints. Given that individual sensitivity to noise and vibration varies significantly, regulations alone cannot fully address N&V issues. If there are numerous complaints from people living near the railway line, the infrastructure owner often has no choice but to address them.

When designing and installing a building or system, engineers should carefully consider the impact of N&V, as the resulting fatigue in structures and systems can shorten their lifespans. Engineers should also account for the comfort of passengers and operation and maintenance personnel. PSD (Platform Screen Door) is a good example of best practices aimed at enhancing ride comfort.

Sources of noise & vibration

The main sources of noise and vibration (N&V) in railway are as follows:

- Friction between the rail head and wheel surfaces of moving trains.
- Air friction along the train car body surface during operation.
- Micro-pressure waves generated when a train enters and exits a tunnel at high speed.
- Noise and vibration from machinery operating in depots.

Noise and vibration are often mentioned together because they originate from the same source – vibration. When vibration occurs, it propagates through a medium like structure, ground, and air. When this transmitted vibration energy reaches our ears within audible frequencies, we perceive it as sound (noise). If the vibration is below the audible frequency, our body perceives it as physical vibration.

Managing noise and vibration in railway projects focuses on mitigating their impact on:

- Internal factors: infrastructure, systems, operation and maintenance personnel, and passengers
- External factors: nearby buildings, systems, and residents

Noise and vibration requirements

The following international standards can be applied:

- WHO Community Guideline: Provides community noise guidelines to protect public health, suggesting the equivalent noise level (Leq) and maximum noise level (Lmax).
- ASHRAE Noise Criteria (NC): Recommends noise criteria for HVAC system in indoor spaces

Although noise limitations vary by country, city, and area, the following general guidelines may be suggested. (To establish N&V requirements, it is essential to study local codes first.):

< Noise around rolling stock >

Area	Operation status	Noise limits
Interior	Stationary	70 dBA (LpAFeq)
	Running 80km/h at open area	80 dBA (LpAFeq)
Exterior	Stationary	72 dBA (LpAFeq)
	Running 80km/h at open area	87 dBA (LpAFeq)

< Noise at local area >

Area	5am to 9am	9am to 6pm	6pm to 10pm	10pm to 5am
An areas within 100 meters from schools, hospital, and home for the aged	45 dB	50 dB	45 dB	40 dB
Residential areas	50 dB	55 dB	50 dB	45 dB
Commercial areas	60 dB	60 dB	60 dB	55 dB
Light industrial areas	65 dB	70 dB	65 dB	55 dB
Heavy industrial areas	70 dB	75 dB	70 dB	65 dB

< Vibration requirements >

- ISO 10137 – Human Exposure Criteria: suggests the effects of periodic vibrations and temporary vibrations within a range of 1 to 80 Hz.
- German DIN 4150 – Structure Damage Criteria: provides reference values.

Noise and vibration management

Railway systems require the integrated noise and vibration management activities across subsystems to ensure that performance and functionality align with stakeholders' expectations. ISO 31000 provides recommended approaches for managing noise and vibration.

External noise and vibration assessment

In the early stage, external noise and vibration should be assessed for the following reasons:

- Establishing current noise and vibration levels provides a baseline for comparing future changes and assessing the impact of railway construction and operation.
- Accurate measurements enable prediction of potential noise and vibration impacts on the surrounding environment and communities, allowing for better planning and mitigation strategies.
- Compliance with environmental regulations and standards requires detailed knowledge of pre-existing conditions.
- Providing transparent information about existing noise and vibration levels helps address community concerns and building trust with stakeholders.
- Identifying areas with high existing noise and vibration levels allows for targeted mitigation measures to minimise additional impacts during and after construction.

Sensitive buildings around the main line should be identified first, followed by an assessment. Typically, sensitive buildings near a railway line are churches, schools, hospitals, etc. If the noise and vibration assessment results indicate that certain areas are likely to exceed the established criteria, mitigation measures should be implemented.

Noise & vibration simulation

To conduct a simulation for noise and vibration, a model needs to be developed using the following input data:

- Infrastructure: Track type (at grade, elevated, or tunnel), track alignment (radius and gradient), speed limit for each section, etc.
- Rolling stock: Axle load, acceleration & deceleration rate, frequency of rolling stock, etc.
- Surrounding topographic map.

The assessor performing the simulation should not rely solely on the simulation results. Small changes in input parameters can lead to significantly different outcomes. It is essential for the assessor to have an intuitive understanding of the expected result and to compare the simulation outcomes with these expectations to ensure their validity and reliability.

Figure 89 - An example of noise simulation

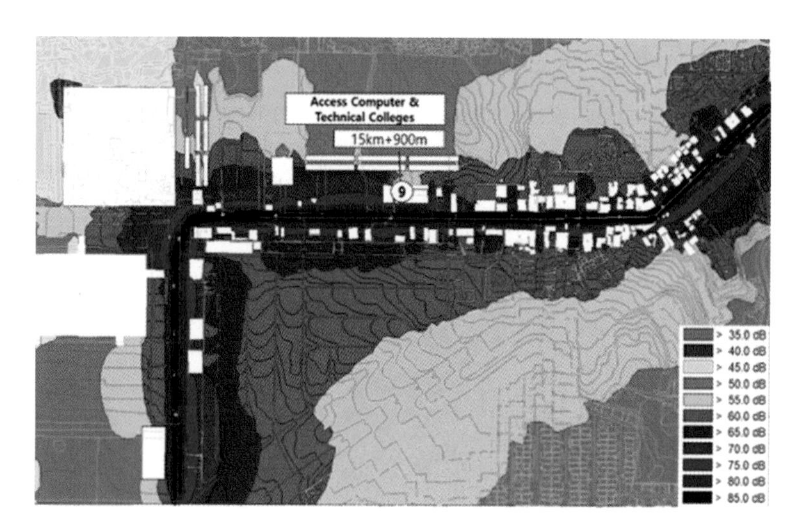

Causes and mitigation measures

In railway projects, the causes and impacts of noise & vibration are generally similar. Figure 90 illustrates the common causes and mitigation measures of noise & vibration, structured in a bowtie format. The blue rectangles represent causes, coded as "C," while the grey rectangles indicate preventive barriers, coded as "PB."

Figure 90 – Causes and mitigation measures of noise & vibration

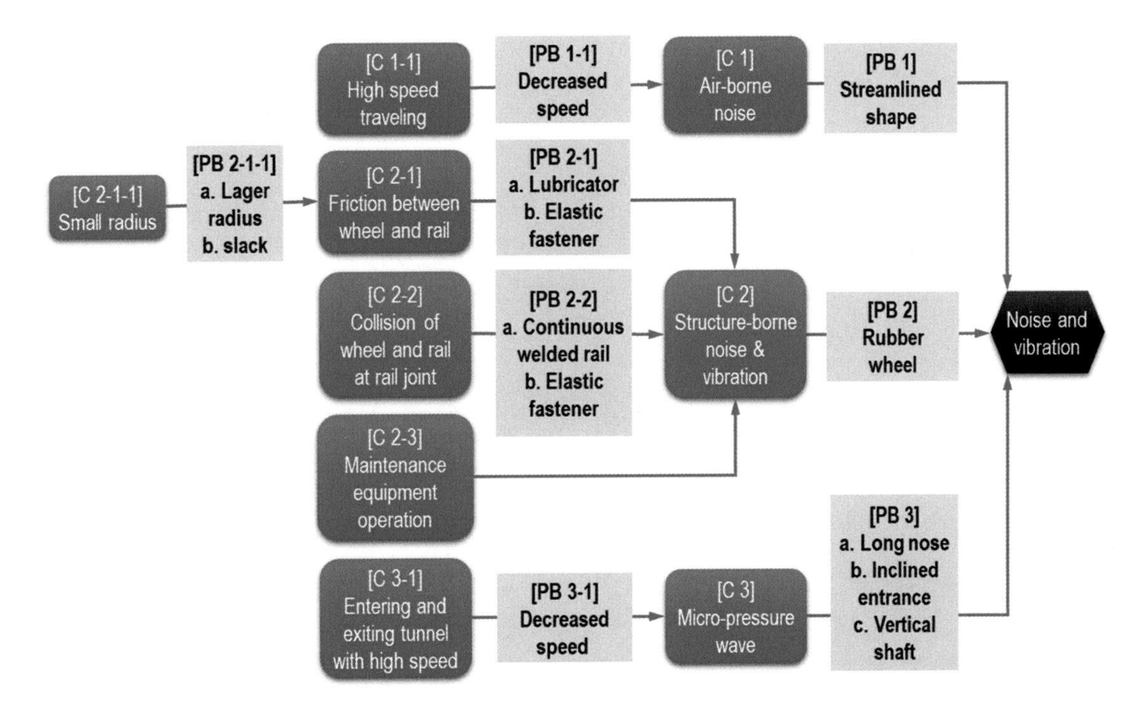

Air-borne noise [C 1]

Air-borne noise in railways arises from friction between the train's car body and the air. As train speed increases, so does the noise level, making this issue particularly significant for high-speed trains. One method to reduce this noise is by decreasing the speed of high-speed trains [PB 1-1]; however, operating at lower speed defeats the purpose of high-speed rail. Therefore, designing a streamlined car-body for trains is advisable. This is why high-speed trains are typically more streamlined than low-speed trains, such as metro trains.

Structure-borne noise and vibration [C 2]

There are three main sources of structure-borne noises and vibrations:

- Friction between the wheel and rail head [C 2-1]
- Collision of the wheel and rail at rail joints [C 2-2]
- Operation of maintenance equipment [C 2-3]

[C 2-1] occurs primarily when a train is travelling at curves, causing friction between the wheel flange and rail head, known as squeal noise, as illustrated in Figure 91. To mitigate noise and vibration, a railway lubricator is used; this device, installed along tracks, automatically sprays lubricating oil on the rail head [PB 2-1-a]. Another measure is to install elastic fasteners, which secure rails to the sleepers while allowing some flexibility [PB 2-1-b]. A small curve radius also contributes to squeal noise [C 2-1-1]. Rails on curves should have slack to reduce impact between the wheel flange and the rail head, allowing slight lateral movement of wheels within the track and minimising force on the rail during curves.

Figure 91 - Wheel and rail

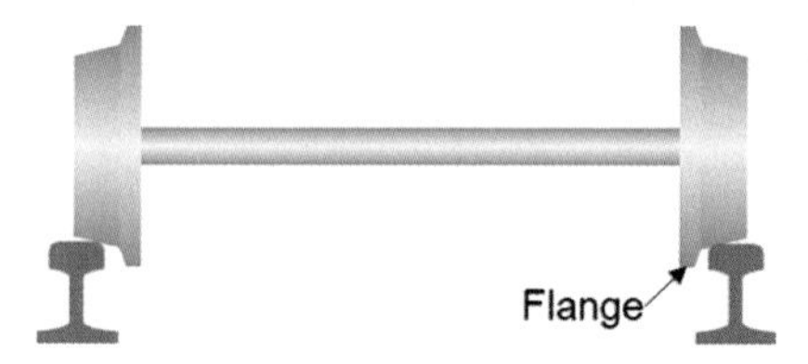

[C 2-2] noise occurs when the wheel hits the rail joint, which is designed to absorb rail expansion due to rising temperatures in summer. To mitigate this noise, continuous welded rails (CWR) should be used [PB 2-2-a]; CWR involves welding long sections of rails into a continuous length, typically without joints. Structure-borne noise & vibration can also be reduced by using rubber wheels instead of steel wheels [PB 2]

Micro-pressure wave [C 3]

A micro-pressure wave (MPW) in a railway tunnel occurs when a high-speed train enters a tunnel, rapidly increasing air pressure at the entrance [C 3]. This pressure wave travels through the tunnel and exits as a

sharp sound or "sonic boom." MPWs result from the compression and displacement of air in the confined tunnel space, and they are more pronounced with higher train speeds and narrower tunnels.
The easiest way to reduce MPW is to reduced speed [PB 3-1]. However, low-speed operation is not fit for purpose of introducing high-speed rail.
A streamlined and elongated train nose gradually compresses the air as the train enters the tunnel, reducing the sudden pressure buildup that leads to micro-pressure waves [PB 3-a]. Bullet-shaped designs are often used, as they smooth air displacement and decrease the intensity of pressure wave. These aerodynamic features are especially crucial for high-speed trains, ensuring a quieter and more efficient tunnel entry.
An inclined tunnel entrance reduces micro-pressure waves by preventing a sudden buildup of pressure [PB 3-b], as shown in Figure 92. This design allows air to be gradually compressed as a high-speed train enters, mitigating the sudden air compression that generates micro-pressure waves.

Figure 92 - Shape of tunnel entrance

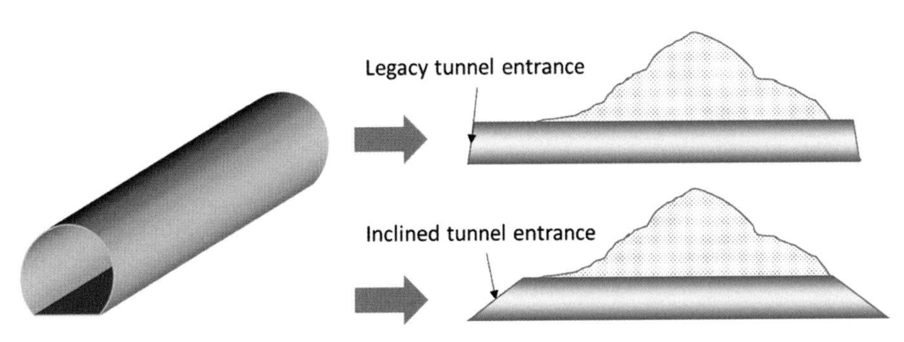

A vertical shaft near the entrance of a tunnel helps reduce micro-pressure waves by providing an outlet for air to escape when a high-speed train enters the tunnel [PB 3-c]. This shaft alleviates the rapid air compression at the tunnel entrance by allowing air to flow upward, thereby reducing pressure buildup inside the tunnel, as shown in Figure 93. By diverting air through the vertical shaft, the intensity of the micro-pressure wave is lowered, resulting in less noise and vibration as the train passes through.

<u>Figure 93 - Vertical shaft</u>

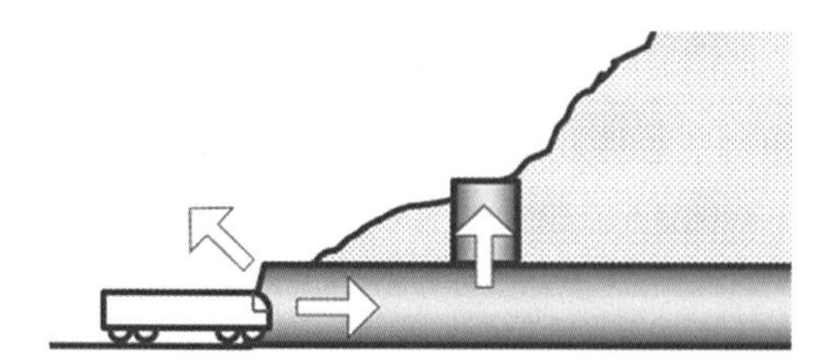

Effects and mitigation measures

Figure 94 outlines the common effects of noise & vibration along with their mitigation measures. The chart uses a bowtie format, where red rectangles represent effects with the code "E," and grey rectangles denote mitigative barriers with the code "MB."

Figure 94 – Effects and mitigation measures

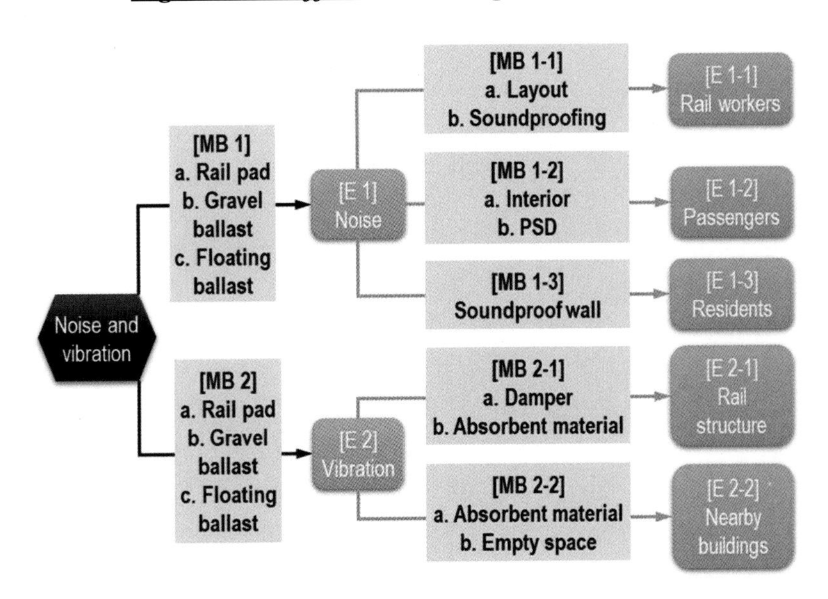

Acoustical effects [E1]

To reduce acoustical effects, the following measures can be applied:

- Rail pad [MB 1-a]
- Gravel ballast [MB 1-b]
- Floating ballast [MB 1-c]

A rail pad is a cushioning material placed between the rail and the sleeper (or tie) [MB 1-a]. It reduces noise and vibration generated by train movement by absorbing and damping the impact and vibrations from the train wheels. This rail pad minimises the transmission of noise and vibration to the surrounding environment and infrastructure, resulting in a quieter and smoother ride.

Gravel ballast is more effective at reducing noise and vibration than concrete ballast [MB 1-b]. The small, loose stones in gravel ballast provide natural cushioning and flexibility, absorbing and dissipating vibrations and impacts from train wheels. This damping effect minimise the transmission of noise and vibrations to the surrounding environment. In contrast, concrete ballast is more rigid and does not offer the same level of cushioning, potentially leading to higher noise and vibration levels.
Floating ballast is a track support system designed to reduce noise and vibration [MB 1-c]. In this system, the track rests on a resilient layer, such as rubber or a specialised composite, which is then supported by traditional ballast. This cushioning layer absorbs and dampens vibrations and noise generated by train movement. By isolating the track from the underlying support structure, floating ballast minimises the impact of noise and vibrations on the surrounding environment and infrastructure.

Rail workers (operation or maintenance staff) are exposed to continuous noise [E 1-1]. Their office room or any location where they spend extended period should be soundproofed and arranged to block the noise caused from train operations [MB 1-1-a, b].

When using trains, passengers mainly stay on the platform and inside of the trains. To reduce passengers' exposure to noise, platform screen doors (PSDs) should be installed on the platform, and the interior of trains should be designed to block the noise from train operations [MB 1-2-a, b].

Soundproof wall is recommended for residents around the line. The height of the soundproof wall should be sufficient to block the noise, as shown in Figure 95.

Figure 95 – Height of soundproof wall

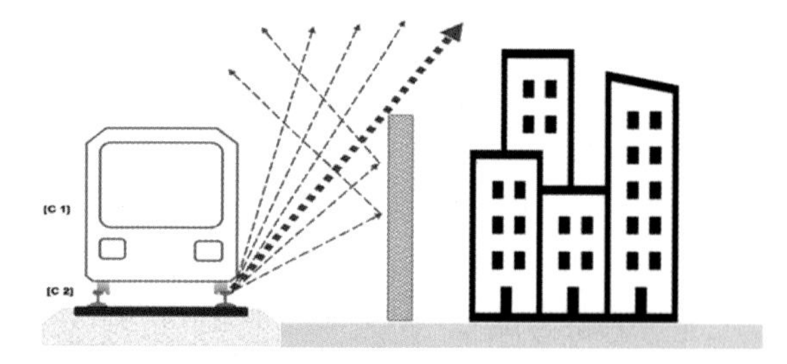

Vibration effects [E 2]

The effect of vibration generally impacts the rail structure and nearby buildings around the line. Since noise and vibration share the same principles of occurrence, the mitigation measures for reducing vibration are the same as those for reducing noise – rail pad, gravel ballast, and floating ballast.

Installing dampers and using absorbent materials are generally suggested to block vibration from rail structures that are vulnerable and near the track [MB 2-1-a, b]. A damper helps reduce vibrations by absorbing and dissipating the energy generated by oscillations. It converts the kinetic energy of the vibration into heat, which decreases the amplitude of the vibration. This leads to improved stability and protects structures or machinery from excessive movement and potential damage.

For nearby buildings, absorbent material can be used [MB 2-2-a]. Absorbent material reduces vibrations by converting the vibration energy into heat through internal friction within the material. This process dampens the vibration, lowering its intensity and preventing excessive oscillation. It helps protect structures or equipment from damage by minimising the transfer of vibration energy.

Sometime designing an empty space between the track and the buildings is recommended [MB 2-2-b]. The empty space can reduce vibrations by providing a gap that prevents the direct transmission of vibration from one surface to another. This isolation limits the transfer of vibrational energy, acting as a buffer that helps reduce the intensity of vibrations reaching other parts of a structure or system.

APPENICES

Acronyms

- AFC: Automated Fare Collector
- ALARP: As Low As Reasonably Practicable
- C/O/M: Construction, Operation, and Maintenance
- CCB: Configuration Control Board / Change Control Board
- CI: Configuration Item
- CM: Configuration Management
- CM: Corrective maintenance
- COTS: Commercial Off-The-Shelf
- CSA: Configuration Status Accounting
- DRACAS: Data Reporting, Analysis and Corrective Action System
- EMC: Electro-Magnetic Compatibility
- EMI: Electro-Magnetic Interference
- EMS: Electro-Magnetic Susceptibility
- EN: European Norm
- FAT: Factory Acceptance Test
- FBD: Function Block Diagram
- FMEA: Failure Mode and Effect Analysis
- FMECA: Failure Mode, Effect and Criticality Analysis
- FRACAS: Failure Reporting, Analysis and Corrective Action System
- FTA: Fault Tree Analysis
- HAZID: Hazard Identification
- HAZOP: Hazard Operability
- HF: Human Factors
- HVAC: Heating, Ventilation, and Air Conditioning
- ICD: Interface Control Document
- ICE: Independent Checking Engineer
- ICNIRP: International Commission on Non-Ionizing Radiation Protection
- IHA: Interface Hazard Analysis
- ISA: Independent Safety Assessment

- IV&V: Independent Verification and Validation
- KPI: Key Performance Indicator
- LBS: Logistic Breakdown Structure
- LRU: Line Replaceable Unit
- MART: Mean Active Repair Time
- MTBF: Mean Time Between Failure
- MTBSAF: Mean Time Between Service Affecting Failures
- MTTR: Mean Time To Repair
- N&V: Noise and Vibration
- OSHA: Operating & Support Hazard Analysis
- P&A: Premises and Assumptions
- PBS: Product Breakdown Structure
- PHA: Preliminary Hazard Analysis
- PM: Preventive maintenance
- PPHPD: People Per Hour Per Direction
- RAM: Reliability, Availability, Maintainability
- RAMS: Reliability, Availability, Maintainability, Safety
- RBD: Reliability Block Diagram
- RTT: Round-Trip Time
- SAP: System Assurance Plan
- SAT: Site Acceptance Test
- SCI: Safety Critical Item
- SCMP: Software Configuration Management Plan
- SE: Systems Engineering
- SEMP: Systems Engineering Management Plan
- SFAIRP: So Far As Is Reasonably Practicable
- SHA: System Hazard Analysis
- SIL: Safety Integrity Level
- SPOF: Single Point Of Failure
- SQAP: Software Quality Assurance Plan
- SSHA: Subsystem Hazard Analysis
- SVVP: Software Verification and Validation Plan

- SWIFT: Structured What-If Technique
- T&C: Testing and Commissioning
- V&V: Verification and Validation
- WBS: Work Breakdown Structure

Standards and References related to Systems Engineering

- IEEE 1220: Standard for application and management of the systems engineering process
- INCOSE Systems Engineering Handbook
- Systems Engineering Body of Knowledge (SEBoK)
- NASA Systems Engineering Handbook
- Systems Engineering Guidebook for Intelligent Transportation Systems
- Systems Engineering Plan (SEP) Outline

Printed in the United States
by Baker & Taylor Publisher Services